The Biology and Management of Animal Welfare

Published by
Whittles Publishing Ltd.,
Dunbeath,
Caithness, KW6 6EG,
Scotland, UK

www.whittlespublishing.com

ISBN 978-184995-366-5

Production managed by Jellyfish Solutions Ltd.

The Biology and Management
of Animal Welfare

FRAUKE OHL, RORY PUTMAN AND MEMBERS OF
DEPARTEMENT DIER IN WETENSCHAP EN MAATSCHAPPIJ
(DWM), UNIVERSITY OF UTRECHT

Whittles Publishing

Contents

Additional box features have been contributed by Nienke Endenburg (pages 17–22), Dominic McCafferty and Rory Putman (pages 72–73)

PREFACE

This book has had a somewhat painful gestation. It was originally conceived of as a joint venture between myself and Professor Frauke Ohl – in part to try and force a break away from some of the current conservatism we perceived in the assessment and management of animal welfare and bring to it a more universal approach based on current understanding of welfare and its biological basis, while also in part designed to serve as a 'primer' in support of the burgeoning number of MSc courses in Animal Welfare appearing in association with veterinary schools and other university departments right across Europe. As such, it is inevitably based in some large part on our own teaching in such MSc courses in the Universities of Utrecht and Glasgow and on our joint research.

As was our established practice in such collaborations, I took responsibility for pulling together a first draft – leaving gaps for Frauke to fill in some text where her expertise was undoubtedly greater than mine, and anticipating critical comments on the outline text and structure I had myself essayed to that point. It is an iterative and interactive process which has served us well in the past. Sadly, while I was working on that initial draft text, Frauke was diagnosed with a particularly aggressive form of cancer and died shortly afterwards; she had just passed her 50th birthday. In her honour I have endeavoured to complete the book we had planned, with input and support from other members of the department she led in Utrecht. I am enormously grateful to them for their generosity in helping me complete the book. Since much of the text is based on ideas we had been discussing for many years, and in large part on previous joint publications in the more academic literature, it remains very much coloured by her ideas and inspiration. We hope the book does her justice and will indeed serve to support Master's courses in Animal Welfare as she had planned.

Frauke Ohl was a remarkable person who made an immense impression on anyone who met her. She was enormously 'driven' – strong, generous and honest – with a tremendous inner strength and certainty deriving from the fact that she was perhaps the most self-aware person I have ever met, strong in that self-knowledge. She was totally committed to behavioural science and its application to improving understanding (and practice) in animal welfare management. Having trained as a researcher at the German Primate Centre in Göttingen and at the Max Planck Institute for Psychiatry in Munich, she was appointed to the position of Professor of Animal Welfare and Laboratory Animal Science in the Veterinary Faculty of the University of Utrecht in 2004, and in 2006 took over as Head of Dier in Wetenschap en Maatschappij (the Department of Animals in Science and Society). Alongside her university work she was very active in extension work, determined to ensure through her own efforts that results of biological understanding were channelled into improving practice in the real world. She served as the President of the Dutch Council on Animal Affairs (Raad voor Dierenaangelegenheden), the Scientific Advisory Committee of the Ministry of Economic Affairs, and as a member of the National Advisory Committee for Animal Experiments. She was, I think, the first international representative on the Board of Trustees of the Universities Federation for Animal Welfare in the UK, an associate member of the European College of Animal Welfare and Behavioural Medicine, and a member of the Advisory Board for the Centre for Alternatives on Animal Testing (Europe). At this same European level of involvement, she was a member of the European Forum of Animal Welfare Councils (EuroFAWC) and of the panel convened by the Federation of European Veterinarians (FVE) tasked with development of a model curriculum on Animal Welfare for veterinarians: 'European veterinary education in animal welfare, science, ethics and law'.

Frauke Ohl was a committed, charismatic and inspirational person. Above all, she was always conscious of the great importance of education, and always enthusiastically supported education within her own faculty and at national and international levels. All royalties from this book will be devoted to a fund set up in the University of Utrecht in her memory to support research in Sustainable Animal Stewardship.

Rory Putman, Banavie, 2017

ACKNOWLEDGEMENTS

RP would like to thank Nienke Endenburg, Franck Meijboom, Jan van der Valk and other colleagues within the Dier in Wetenschap en Maatschappij of the University of Utrecht for their assistance and support in helping me to complete this book

We are grateful also to the Dutch Council for Animal Affairs (Raad voor Dierenaangelegenheden) for permission to reproduce Figure 6.1 and other concepts explored in Chapter 6 from earlier publication in:

> Raad voor Dierenaangelegenheden (RDA) (2010) *Moral Issues and Public Policy on Animals.* Report 2010/02. s'Gravenhage, The Netherlands.
> Raad voor Dierenaangelegenheden RDA) (2012) *Duty of Care: Naturally Considered.* Report 2012/02. s'Gravenhage, The Netherlands.

Much of the discussion of how one might extend considerations of welfare and its assessment from the individual to the group (see Chapter 5) has been presented in a previous publication:

> Ohl, F. and Putman, R.J. (2014) Animal welfare at the group level: more than the sum of individual welfare? *Acta Biotheoretica* 62(1), 35–45.

We are also very grateful to the publishers of the journal *Biotheoretica* for allowing us to draw freely on that material again here.

1

A WORD IN ADVANCE

In 1965 the Brambell Committee[1] reported to the UK Government on their investigation of 'the welfare of animals kept under intensive livestock husbandry systems'. The report had far-reaching implications and reflected a growing concern about our responsibilities and the way we treat and manage animals. Even at that time, the Brambell Committee noted that,

> 'Welfare is a wide term that embraces both the physical and the mental well-being of the animal. Any attempt to evaluate welfare, therefore, must take into account the scientific evidence available concerning the feelings of animals that can be derived from their structure and function and also from their behaviour'.

Despite this wider definition, the main part of the report restricted itself largely to recommendations that would seek to implement changes in management, designed explicitly to prevent suffering in intensively managed livestock. Fair enough: that was their remit.

But welfare science and our understanding of the factors affecting the welfare of individual animals or larger groups has come a long way since these early efforts at improving the lot of animals in our direct care. Most recent analysts now acknowledge that at any one point in time, an individual animal's welfare status lies on the continuum between negative/bad welfare and positive/good welfare. Continued adherence to the Brambellian focus on the simple avoidance of negative states thus masks consideration of those factors which may help promote positive welfare as well as avoidance of negative welfare states.

---

1 Brambell Committee: a technical committee set up by the UK Government in 1965 to inquire into the welfare of animals kept under intensive livestock husbandry systems (Brambell Committee (Report), HC Deb 15 December 1965 vol 722 cc279-80W).

More fundamentally, a view of welfare which is dominated by an emphasis on the avoidance of negative states neglects the fact that – except in the specific instances where natural selection processes have been largely countermanded by deliberate selection by humans – animals have evolved, optimising the ability to interact with and adapt to (changing conditions within) their environment and that their exposure to environmental challenge and short periods of 'negative welfare' may be inevitable if these are understood as triggers to release from the animal's repertoire the appropriate behavioural or physiological response to adapt to those challenges. Although it has taken some time for this idea to be more generally adopted, and it is by no means universally reflected in the literature, many now do advocate this more dynamic view of welfare, such that a welfare issue arises only when an animal or a group of animals have insufficient opportunity (freedom) to respond appropriately to a potential welfare 'challenge' through adaptation by changes in its own behaviour.

On this basis we may then suggest that assessment of welfare should therefore focus not so much on the challenges that any animal may face at a given moment but rather on whether or not the animal has the freedom and capacity to react appropriately (i.e. adaptively) to both positive and potentially harmful (negative) stimuli. By the same token, welfare should not be considered as an instantaneous construct to be assessed at some moment in time. An adaptive response may take some finite period of time; crucially, therefore, our assessment of welfare should not simply consider the status of any individual at a given moment in time, but needs to be integrated over the longer time periods required to execute such change. A further problem implicit in standard methods for objective assessment of welfare status is that such protocols inevitably reflect the observer's perspective and subjective judgement, whereas most modern commentators would now acknowledge that, to some significant degree, any animal's status must be that perceived and judged by that animal itself.

Such review suggests that instead of considering individual welfare in terms of some 'universal' or 'objective' state as might be assessed by an external observer, to the animal itself, its welfare status is a function of a subjective self-evaluation or self-perception. And increasingly we begin to understand the physiological and neural processes which underpin an animal's ability to assess and respond to its own sense of 'well-being'.

Despite these advances in our scientific understanding, approaches and attitudes to welfare and its management have changed remarkably little in 50 years. The report of the Brambell Committee has cast a long shadow over welfare practice, which still tends to focus on avoidance of suffering in management, and depends on the same Five Freedoms first proposed by the Brambell Committee, with a presumption that animal welfare is preserved if the animals are kept free from:

◊ hunger, thirst or inadequate food,

◊ thermal and physical discomfort,

◊ injuries or diseases,

◊ fear and chronic stress, and

◊ are free to display normal, species-specific behavioural patterns.

The first four of the five *freedoms* were formulated from the perspective that the absence of actual negative impact assures welfare; only the fifth, although more indirectly, potentially implied an expectation of facilitation of more positive aspects of welfare.

As we shall explore later in the book, there have been various reformulations of these essential principles, but these different incarnations change the basis of the construct little and, as an instant index, the Five Freedoms remain widely used today as a guideline for welfare assessment protocols, with the actual state of welfare of an animal being characterised as unimpaired if it complies with those five freedoms.

In large part, this is perhaps a direct consequence of context: the Brambell Committee's report to Government was specifically in investigation of the welfare of intensively managed livestock – and indeed much of the current work on welfare relates to closely managed animals, whose entire environment is controlled by human agency and whose welfare likewise can thus be managed by changes to that controlled environment. In consequence, assurance of the Five Freedoms (or some variant of those) may indeed assure adequate welfare (avoidance of suffering), even if it does not actively promote positive welfare. Promotion of positive welfare may be considered, in terms perhaps of projects of environmental enrichment for enclosed animals (be they livestock, laboratory animals or captive zoo animals), but these are still explored within the primary framework of ensuring the five freedoms, in that in effect they seek to ensure provision of the fifth freedom: the opportunity to 'display normal, species-specific behavioural patterns'.

What is disappointing in this regard is that it is clear that the Brambell Committee's report never set out to be a 'welfare concept', but was developed specifically to establish minimum requirements to ensure the absence of negative welfare. Despite this – and despite the Brambell Committee's own broader definition of what is meant by welfare (presented in our very first paragraph here) – many, over the years, have taken the Five Freedoms to define what is implied by welfare itself, and taken their provisions to be necessary and sufficient to ensure positive welfare. However, in its strict adherence to this approach, welfare management becomes very rigid in its approach, very inflexible and actually takes little note of the very considerable advances which have been made in the intervening years in our understanding of the actual biology of what constitutes and contributes to an animal's welfare. Inertia is further encouraged by the very compartmentalised nature of existing welfare legislation, such that there may be a clear legal distinction between responsibilities defined towards farm animals, laboratory animals, companion animals, captive wild animals and free-ranging wildlife, making it more difficult to establish common principles which cut across such categorisation.

We have an enormous admiration and respect for the achievements of the Brambell Committee in improving welfare and attitudes to welfare amongst managers of closely managed animals. Given the attitudes and practices of its time, the scale of improvements delivered in response to their report in the care of animals, and particularly intensively managed farm livestock, was colossal. But we suggest that entrenchment of that same approach in our present-day discussions about welfare and its management, ignoring subsequent advances in our knowledge, is more of an affront than it is a compliment to those early pioneers of animal welfare management. In this book we offer a personal review of some of those advances in our understanding of the science of welfare and welfare management, in an attempt to encourage readers to 'think outside the box' of current welfare management and develop new approaches to the measurement and management of welfare which may be extended more widely and with greater flexibility.

It was, in fact, the need to respond to a request to extend existing approaches for the assessment and management of welfare to free-ranging wild animals (Ohl and Putman, 2013a, 2013b, 2013c; Ohl and Putman, 2014b), that made us appreciate that the general principles and protocols developed for more closely-managed animals simply could not be applied in more uncontrolled environments.

It was that realisation that forced us to return to first principles and explore the biological fundamentals of what constitutes and contributes to welfare, working from that basis to consider how that biological understanding might help underpin more functional ways of assessing welfare and addressing welfare problems. And if those approaches are developed from the fundamental biology of what may result in good or poor welfare in *any* animal, then we should be able to develop approaches to the assessment and management of welfare that might be applied, irrespective of context, to closely managed animals or those less closely managed, and whether they are farm livestock, lab animals, pets or pests. After all, as Webster points out, whether it is a child's pet, an experimental animal or a pest, 'a rat is still a rat' (Webster, 1994).

At the same time, as scientists, we recognise that whatever our understanding of the underlying biology, any objectivity in analysis must inevitably cede to the subjectivity of ethical assessment when determining whether any given welfare status is or is not 'acceptable' to society. The 'translation' of welfare assessments into management practice and the way in which that management practice is viewed by society more widely are markedly affected by public understanding and public attitudes. In this book, we will therefore explore how biological understanding and an understanding of ethical values of society must be integrated into any 'universal' framework for welfare management.

In recent years, animal welfare issues have gained an increasingly prominent place in public debates. However, there is no consensus on how to measure the welfare status of an animal objectively or how to assess the welfare implications of any given management practice. At the same time, it is clear that every definition of animal welfare is influenced by the moral or ethical standards of society and that not only are attitudes affected by the cultural context and traditions of a given society but that there is also a growing diversity of views on animals, even within any societal grouping. Furthermore, an inconsistency can be observed between the importance that people say they attribute to animal welfare and the things they actually do, directly and indirectly, for animal welfare. In order to tackle some of today's animal-related issues, it is therefore necessary to answer some underlying fundamental questions about human attitudes and human expectations. But in raising the question of moral stance, we should close this introduction by making it clear that, in our treatment, we draw a clear distinction between animal welfare and animal rights. Our considerations in the following chapters relate only to responsibilities in relation to **animal welfare**; we do not in any way address the rather separate issue of **animal rights** (e.g. Singer, 1989; de Fontenay, 2006; Haynes, 2011; *inter alia*). Thus, quite explicitly and quite deliberately, in this book we do not address questions of whether or not humans have the right to exploit animals for food, to use them as laboratory models, to hunt, or to keep animals as pets. We simply consider what may be the duty of care and requirements of action to ensure acceptable welfare of wild or more closely managed animals, whatever the (philosophical) debate about rights and wrongs of management in the first place.

Definitions of welfare and welfare states

Perhaps before we attempt to go any further we should explore exactly what it is that we mean by welfare? Welfare is in fact relatively rarely explicitly defined; rather, there is a general presumption that we implicitly understand what is meant by the term (as also the term 'well-being'). When we do search in the literature we find a plethora of rather vague encapsulations.

Broom (1988) defined welfare as 'an individual's state as regards its attempts to cope with its environment', while noting that 'feelings, such as pain, fear and the various forms of pleasure, are a key part of welfare'. Duncan (1993) and Fraser and Duncan (1998) suggest that welfare is entirely to do with how animals feel. At the same time, those with a medical or veterinary background sometimes present the view that physical health is all, or almost all, of welfare.

M.S. Dawkins (1990) stated that 'the feelings of the individual are the central issue in welfare but other aspects such as the health of that individual are also important'. Sejian *et al.*, (2011) note that welfare may be considered as 'the ability of an animal to cope physiologically, behaviourally, cognitively and emotionally with its physiochemical and social environment', while Webster (2012) notes that, 'There is now broad agreement amongst academics and real people that the welfare of a sentient animal is defined by how well it feels; how well it is able to cope with the physical and emotional challenges to which it is exposed'.

In a similar way, we may explore various definitions that have been offered in relation to suffering. Fraser and Duncan (1988) denote suffering as 'strong, negative affective states such as severe hunger, pain, or fear'. Dawkins (e.g. 1990, 2008) suggests that 'suffering can result from experiencing a wide range of unpleasant emotional states such as fear, boredom, pain, and hunger'. By converse, Appleby and Sandøe (2002) note that 'Animals should *feel* well by being free from prolonged and intense fear, pain and other negative states, and by experiencing normal pleasures'. Suffering thus describes the negative emotional experience resulting from being exposed to an acute or prolonged state of negative welfare.

As we progress through the book we will, to some extent, refine these descriptions and try to offer more specific definitions for what we mean by welfare and different welfare states. However, it remains one of the problems of any new area of enquiry – and the welfare 'area' is no exception – that successive authors tend to adopt words from common parlance and give them quite technical meanings. This can actually lead to misunderstanding when a reader's understanding of a word, which is now used in a precise and specific way, is coloured by familiarity with its original non-technical overtones. To avoid any such misunderstandings and potential misinterpretation, we therefore offer a glossary at the end of the volume to make clear what we mean by each term when used.

2

ETHICS AND THE CHANGING ATTITUDES TOWARDS ANIMALS AND THEIR WELFARE

FRANCK L.B. MEIJBOOM

Animal welfare: more attention, but no consensus

Discussions on the welfare implications of the practice of tail docking of pigs, the welfare of pedigree dogs or of stranded whales. These are just a few examples of the increase of public attention for animal welfare. Independent of any question of whether humans have the right to use and exploit animals in the first place, the primary debate focuses on public demands to minimise animal suffering and optimise animal welfare in managed populations of animals. This attention to animal welfare is not limited to the level of personal views and opinions or restricted to the Western world only. Animal welfare is on the public agenda. It is implemented in national and European laws and regulations (e.g. in the Lisbon Treaty of 2009 and Directive 2010/63/EU, art. 12), and often can be recognised in politics and public debate.

In spite of all this attention, the discussions on animal welfare are characterised by fundamental differences and apparent inconsistencies. Although mice in animal experiments in Europe by law have to be closely monitored in terms of animal welfare, the welfare of mice kept for company is often not regulated. Similarly, while unsedated slaughter of mammals often raises fierce debates on animal welfare, the necessity of sedation in killing fish is less often debated. First, these differences can be partly explained by changes in scientific knowledge. There is no consensus on how to measure the welfare status of an animal objectively or the welfare implications of any given management practice. At the same time, science is progressing in this area. For instance, a few decades ago the scientific evidence that fish could experience pain was less conclusive than it is today (Braithwaite, 2010). Second, it can be partly a lack of knowledge at the level of animal owners or the general public: for instance owners of obese cats are often convinced that their animal is well cared for and they have the idea that their animal does not have any serious welfare problem.

Third, these differences can be partly explained from a sociological and psychological perspective. Views on animals in general are indisputably culture-, time-, place- and context-dependent. I perceive and value my own dog differently to stray dogs in Romania and, from a Northern European perspective, bull fighting in Spain is evidently an animal welfare problem that should be banned, while a Spanish colleague may consider it less problematic and question instead the welfare of semi-wild animals in the Netherlands (Chapter 6). At the same time, people are often inconsistent in their views about animals: one and the same person might, depending on the circumstances, consider individual housing of rabbits unproblematic in his own backyard, while signing a petition to ban it in animal testing.

These explanations help to understand the actual differences in the animal welfare debate, but do not explain whether the plurality of inconsistencies as such often is considered problematic. In many other contexts, such as music or education, we have many and sometimes mutually exclusive views on what is good music or education. Therefore, the question is what makes animal welfare different? To answer that question and the question of whether differences in animal welfare are problematic, we have to address the ethical dimension of animal welfare. This ethical dimension shows that animal welfare is special: it is not about facts and figures only, but is intrinsically related to values and ethical principles. Science can answer questions of 'what is', but to ultimately answer questions concerning animal welfare one must also deliberate and answer questions about 'what ought to be'. Thus, animal welfare discussions are not limited to the question of what is constitutive for the welfare of the animal, but really begin with public views on why animals are morally important and why animals' interests should be taken into consideration. Answering the latter question is not only relevant for philosophers, but is also necessary to deal with the practical dilemmas of implementing animal welfare.

This chapter aims to explore the ethical dimension and show how the relative recent development of including animals in the moral community is still an ongoing process and underlies many of the current animal welfare debates. Consequently, knowledge of the ethical views on animals and welfare is of direct relevance for dealing with daily questions of animal welfare.

From 'just' animals to questions of justification

Animals have received a lot of attention in history. They have been perceived as (worldly images of) gods, such as in ancient Egypt, or have themselves been considered to be holy or unclean, such as cows or dogs in India.

Nonetheless, animals have mostly been at the background of Western thought. In most cases Schopenhauer's expression seems to be appropriate that at the entrance of Western ethics there was a sign that reads: 'Animals should stay outside' (Schopenhauer, 1840). Traditionally only humans were considered to possess moral status, i.e. only to humans can we have direct duties. Even Jeremy Bentham who has become famous for claiming that '...the question is not, Can they reason? nor, Can they talk? but, Can they suffer?' relates this claim to animals in a footnote to his *Introduction to the Principles of Morals and Legislation* (Bentham, 1789: Chapter 17). This position of the animal in the footnotes of Western philosophy remains up until recently. Even comparatively modern philosophers like John Rawls or Thomas Scanlon have not much to say about animals. Their contractual accounts do not fully exclude the idea that animals can be morally relevant, but mainly focus on the morality of what we owe to human persons (cf. Scanlon, 1998: 179). This position that animals do not directly belong to our moral community, however, does not mean that we should not have concern or care about animals. But, the reason why I should care for a dog starts in my obligation towards other humans rather than the dog itself: I should be careful with the dog of my neighbour, because it is her property.

However, for most, this view on the animal as an outsider to our moral circle has changed. Since the second half of the 20th Century, there has been an increasing public and academic attention to the question of what we owe to animals. More importantly, there is a clear shift in the reason why this is a relevant question. Since the 19th Century, the question of what we might owe to animals had been discussed widely. The cruelty and maltreatment of animals was considered more as a public problem. As a consequence, organisations were established that aimed to improve the treatment of animals, such as the establishment in the UK of the Royal Society for the Prevention of Cruelty to Animals (RSPCA) in 1824, the Dutch Society for the Protection of Animals in 1864 and the German 'Tierschutzbund' in 1881. Although these societies genuinely care for animals, the reason for their establishment mainly started because of concerns about the demoralization of civilization that comes – or was claimed to have come – with cruelty to animals (e.g. Boddice, 2009). This implies that one has strong motives to care for the animals and their welfare, but the reason why one should do so is still embedded in considerations of duties to other humans. From this perspective, cruelty to animals is not so much a problem because of duties towards the animal, but because it does not fit to our human morality. It is inhumane to treat animals in a way that does not consider their welfare. This attitude resulted in a lot of improvements to the living conditions of animals, but from this perspective, animals are still not necessarily granted with status in their own right and were still considered as outside a largely human moral community (Manning and Serpell, 1994; Armstrong and Botzler, 2003).

This view, however, has been radically challenged since the 1970s. Authors such as Ryder (1970, 1971), Singer (1975), Callicott (1980), Rollin (1981), Midgley (1983) and Regan (1983) made it clear that, in their view, animals were no longer considered as mere things that can be used for humans aims, but as entities that have independent interests and belong to the moral community. As a result, at least in the view of these commentators, an animal was seen to have independent moral status and humans might therefore be considered to have direct duties to them. In practice this implies that cruelty to a cow is not a moral wrong simply because it is a waste of resources or because it is a sign of bad moral character, but because the moral position of the cow itself is sufficient reason to care for its welfare. Consequently, any negative impact on animal welfare asks for a moral justification.

Animal welfare and the plurality of views on the moral status of animals

In spite of the increase of academic and public attention to animals and their moral status, there is still a lot of debate about whether and to what extent animals are moral entities in their own right – and not all schools of thought would accept this as a valid assertion. Indeed, we may agree with Fraser (2003) who claimed that debate is 'perhaps too mild a term. The treatment and use of animals has long been a subject of passionate disagreement in Western culture'. On the one hand, animals have a special position in Western society and human–animal interactions are considered to be important and valuable (cf. Herzog, 2011). On the other hand, there is no single relationship between humans and animals. Humans keep animals in many ways and for very different purposes. In some cases, animals are treated as members of family and valued for their own sake. In other cases, the emphasis is still mainly on their instrumental value, or the animal in question may even be considered as a pest. Both types of relationship may imply a strict distinction between humans and animals. This plurality crosses all contexts and cannot be framed easily in terms of the distinction between pets and production animals or between kept and (semi)-wild animals. This poses problems as to whether or not our attitudes or obligations to animal welfare may differ in different contexts of engagement – a construct explored further in Chapter 7.

This diversity reflects the plurality of views on the moral importance of an animal. We lack one Archimedean point, a publicly agreed-upon endpoint of debate. Despite the overview offered earlier, based on both the academic literature and the public debate (European Union, 2016) on the ethics of animal use, it is possible to draft a continuum of views on position of animals ranging from the idea that an animal has mainly instrumental value to the perspective on animals having an inherent worth equivalent to humans.

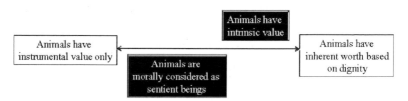

Figure 2.1 *The continuum of views on the moral value of animals. The views included in the figure represent the main positions, but should not be perceived as a comprehensive list.*

The relevance of the continuum of views in Figure 2.1 is not restricted to mapping the fact that this plurality exists – it also helps in understanding and getting a grip on wider discussions on animal welfare. Interestingly, from whatever position, all traditions of thought can consider animal welfare as an important criterion within our interaction with animals. This may suggest that – although we lack one common perspective – animal welfare can function as the overarching concept in the discussion on the human–animal interaction. However, before jumping to this conclusion, it is important to reflect on the differences between the views that allow room for, or even actively use, animal welfare.

Starting with those who consider animals as mere instruments, animal welfare can easily be considered as a relevant criterion. One sometimes has strong practical reasons to care for the welfare of the animal because, for example, a dog may turn aggressive or a cow may not grow or produce milk to maximum efficiency. In this view, animal welfare is important as long as it serves our human interest. This often goes with a definition of animal welfare that mainly focuses on aspects of biological functioning, so that, for example, welfare is assessed as positive if an animal reproduces or grows (cf. Fraser *et al.*, 1997).

For those who consider animals to be morally important for their own sake because they have the capacity to experience pain and pleasure, animal welfare becomes a central criterion to evaluate an animal practice from a moral perspective. With the recognition of an animal as a sentient being, which can experience hurt, there follows a moral duty to care for the animal's welfare. As a consequence, attention to animal welfare is no longer dependent on human interest only. Such a view often goes with a definition of animal welfare that takes a feelings-based approach and provides room for both negative and positive emotions. In practice this implies that improving welfare includes creating living conditions that an animal experiences as positive. In combination with a more utilitarian style of moral reasoning, this view can result in the position that animal welfare is not just an important criterion to ethically assess our interactions with animals, but also the only legitimate argument.

Next to the 'animal welfare only' view, we find other positions that agree on the importance of animal welfare, but stress the need for a broader moral vocabulary than the utilitarian view: for instance, views that argue that any recognition of an animal as a sentient being not only leads to a duty not to harm animals, but leads to the principle of showing active respect to the intrinsic value of that animal. Although this concept of intrinsic value is not self-evident in its scope and normative power, it implies that, based on some (cognitive) capacities, ethics should focus on the individual and, in acknowledgement of this, each action should show respect to the individual animal. This can be reflected in care for animal welfare, but also touches upon a range of new ethical considerations such as animal integrity (cf. Rutgers and Heeger, 1999). The latter may imply that if tail cutting of pigs for production reasons would be positive in terms of animal welfare, such a procedure may nonetheless not be appropriate because it implies human adaptation of the animal to its instrumental function rather than a respect for it as a creature with its own moral standing. The role of these additional arguments next to animal welfare becomes even more prominent if one shifts to the view that animals have inherent worth and dignity and therefore have rights. From an animal rights perspective, attention to animal welfare is important, but even a perfect state of welfare can never be a justification of an infringement of a right, such as the right to freedom or life.

The aim of presenting this continuum is not to label persons or groups, but to stress that one must be aware that starting from different stances on the moral continuum has practical implications for the consequential responsibilities in relation to welfare. Depending on one's position on the continuum of views on the moral position of the animal, one will inevitably formulate different duties towards an animal. Consequently, the strength of

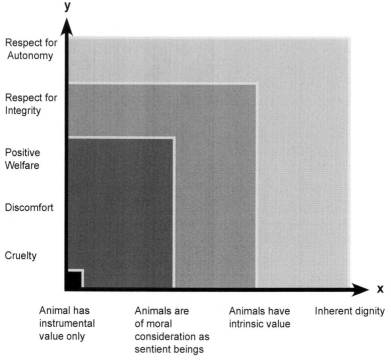

Figure 2.2 *x-axis: The continuum of views on the moral value of animals, y-axis: The resultant range of moral duties towards animals*

the moral imperative to pay attention to animal welfare differs from one as a form of supererogation to one out of a set of many direct duties.

To illustrate this special position of animal welfare, let us take an example of two persons who both claim to care for laying hens. Furthermore, they both recognise that animal welfare is important and they both have equal expertise and knowledge. Nonetheless, they come to completely different assessments of the poultry welfare. Suppose the first person is someone who considers poultry as useful instruments that should be handled with care, otherwise they might not produce in the most optimal way, which is undesirable given the scarcity of natural resources and an objective of maximising production in a sustainable way. He understands animal welfare as balanced biological functioning. The second person considers chickens to be morally relevant for their own sake and refers to animal welfare in terms of positive emotions and natural behaviour. If these two persons assess a housing system in which laying hens biologically function very well, but appear to be unable take dust baths and have trimmed beaks due to problems with feather pecking, it is likely that they will reach opposite judgements about the welfare of the animals housed in the system. This controversy cannot be simply solved by better or more welfare research because it starts in their different views on why animals are morally important, what we owe them and how important welfare is. If this plurality is disregarded, it has direct practical consequences and generates serious confusion.

The limits of a key concept

At this point we can conclude that in the context of animal welfare, ethics and science always go together (see also Chapter 7). Speaking about animal welfare continuously refers to ethical assumptions about the moral position of the animal and the moral relevance of welfare, just as developing an ethical framework for animal welfare would be inappropriate without taking into consideration what we empirically know about what constitutes animal welfare in practical, biological terms.

Furthermore, it is important to conclude that the plurality of ethical views on the moral position of animals has a direct impact on the animal welfare debate. Although there are clear trends, we still lack one common shared moral starting point. At the same time, it is undesirable to disregard the plurality of views to suggest that in practice we all agree on the content and importance of the concept of animal welfare and its importance. If we would do so and then take the view that animal welfare can serve as the overarching concept in the public discussions on animal use, we run into the problem that we pay simultaneously too much and too little attention to animal welfare. On the one hand, we do not pay enough attention because animal welfare as the 'lingua franca' of the public debate invites people to translate all kinds of considerations in terms of animal welfare. This may not necessarily be a deliberate choice, but a consequence of the fact that because animal welfare arguments are considered as the legitimate point of reference, people start to (re)phrase their concerns as animal welfare concerns. For instance, in the debate on the early separation of cow and calf, implicit relational-based or rights-based arguments are put forward as animal welfare concerns. As a consequence, the debate on animal welfare has to cope with such a variety of questions that it hampers the conceptual scientific discussion. All of a sudden animal welfare scientists need to answer questions which actually arise from public views on sustainable animal farming rather than on the long-term effects on social behaviour. This can easily damage the debate and can distract the attention.

On the other hand, establishing animal welfare as the overarching concept may result in society paying too much attention to this concept. A strict focus on animal welfare would imply that important ethical issues remain unaddressed. At first sight this claim may seem an exaggeration: a number of questions and problems with respect to animal use that go beyond animal welfare can still be framed in terms of animal welfare. This holds, for example, for discussions on domestication or the aims for which animals are kept. All appear to perfectly fit into an animal welfare discussion; however, public discussions on tail docking or keeping animals in circuses show that if these topics are only considered from an animal welfare perspective, we do not see the full picture. For instance, for those who start from the view of an animal as a being with intrinsic value, the question whether animals can be kept in a circus is not only about animal welfare. Even if one could argue that the welfare of an animal is not in danger, it still leaves the question open about whether the objective of human entertainment is a sufficient justification to keep and use an animal. This touches upon the distinction between arguments pertaining to animal rights as opposed to those relating to welfare, touched upon in the previous chapter (Regan, 1983; Singer, 1995). However, it is apparent from Figure 2.2 that even if one might deny the concept of animals having rights, if one nonetheless acknowledges that an animal has intrinsic value, the debate on what we owe to an animal cannot be limited to welfare arguments only. Therefore, animal welfare can only function as key concept in the debates on animal use if we are aware of its limits.

Conclusion

At the end of this chapter, we can conclude that for an evidence-based and ethically informed debate on animal welfare one first has to be aware of the multi-layered character of animal welfare, including its ethical dimension. The animal scientist or veterinarian cannot escape the ethical dimensions of animal welfare, just like the ethicist or social scientist cannot ignore the biological or veterinary complexity of this concept. In addition, they all have to deal with the fact that animal welfare is more and more a public theme that is regulated by (inter)national laws and regulations.

Second, it is important to be aware of one's ethical position and that of others with regard to the moral status of animals. It is crucial to be aware of the range of moral stances and the related ideas about our moral duties towards animals. This enables to better understand differences in welfare definitions and in the assessment of available evidence.

Finally, it is important to be aware of the limits of animal welfare. On the one hand, animal welfare as the 'lingua franca' of the public debate invites people to translate all kinds of considerations in terms of animal welfare. As a consequence, the debate on animal welfare has to cope with a variety of questions that its hampers the conceptual scientific discussion. On the other hand, it is important to be aware that sharing an understanding of the moral importance of animal welfare still leaves room for plurality of view on the moral status of animals and what we owe to them.

ATTITUDES TO ANIMALS: THE PSYCHOLOGY OF CONCERN FOR ANIMAL WELFARE

Nienke Endenburg

Welfare is a field that attempts to confront some of the most contentious and important questions of ethics, social science, production, health, economics and politics (Fraser, 2008). In addition to the fundamental position of the ethical values of individuals and wider society in determining attitudes to animals and attitudes to animal welfare, considered in the previous pages, there are also important factors derived from the welfare and underlying psychology of the human individual.

There is a growing recognition that there is a strong interconnection between the welfare of humans and their animals – and not just in the methodologies used to assess welfare or the external factors which may affect welfare status; animal welfare specialists have long noted situations in which animal welfare is directly tied to human welfare (Colonius and Earley, 2013). This interconnectedness between human and animal welfare extends well beyond a commonality in the underlying conceptual or physiological frameworks or in relevant research on the biology and management of welfare in humans or other creatures, but it represents a genuine interaction.

It has long been established that caring for animals may have a very positive effect on human welfare and well-being, while poor welfare of human owners (whether through ill-health, poverty or other circumstances) may be reflected in poor welfare of the animals in their care (Scarlett *et al.*, 1999; McGreevy *et al.*, 2005; German, 2006; Harding, 2014).

Pets can provide social support to their owners (Serpell, 2002), and in this way can improve the mental and physical health of people. The first evidence of an association between the ownership of pets and human health was in the 1980s in which, among a group of victims of heart attacks, survival rates of those owning pets was significantly higher than that of non-pet owners (28% of pet owners survived for at least a year as compared to 6% of non-pet owners; Friedmann *et al.*, 1980). A later study with hypertensive stockbrokers were assigned randomly to either pet or non-pet conditions. Six months later, when put in a stressful situation, subjects in the pet group showed lower increases in blood pressure than those in the non-pet control condition (Allen *et al.*, 2001). Besides its effects on physical health, research has also shown that interaction with a pet animal will have psychological benefits. These include studies showing that pet owners have higher self-esteem, more positive moods, more ambition, greater life satisfaction and lower levels of loneliness (El-Alayli *et al.*, 2006).

Recognition of the therapeutic effects of caring for animals has led subsequently to various forms of Animal Assisted Interventions. These can be rather general in nature where patients, particularly the lonely and the elderly, may benefit from having regular visits from animals they can pet and stroke (PAT dogs). But animals may also be involved in more specific therapies where an approved treatment provider guides interactions between a patient and an animal to realise specific goals (Chandler, 2005). This is not a stand-alone therapy, but has been applied to a wide variety of clinical problems like hypertension (Handlin *et al.*, 2011), stress reduction (Havener *et al.*, 2001), reduced anxiety when people were hospitalised for heart failure (Cole *et al.*, 2007), compromised mental functioning, like senile dementia (Kanamori *et al.*, 2001), emotional difficulties, including autism (Barker and Dawson, 1998), and behavioural distress (Nagengast *et al.*, 1997). Assistance dogs (AD) is a general term that refers to dogs that are specially trained to assist individuals with disabilities. The most common type of assistance dogs is guide dogs (GD) trained to help individuals who are blind or vision impaired. But there are two other types of ADs: service dogs (SD), which assist people with mobility impairments, and hearing dogs (HD), which assist individuals who are deaf or hard of hearing. Studies have demonstrated that 'seizure-alert dog' owners with epilepsy exhibit improvement in seizure rates. One of the most difficult aspects for patients with epilepsy is the unpredictability of seizures (Ortiz and Liporace, 2005).

Equally, the converse is true: where humans are in poor welfare status – perhaps because of poverty, ill-health, malnutrition, ignorance – this is often reflected in poor welfare of their animals. There are owners who, sometimes due to circumstances, do not have the financial resources to give their animal(s) the veterinary care or appropriate food. It can be that they lost their job, or aren't able to work due to poor health. This can lead to compromised animal welfare and sometimes also a reason for relinquishment of the animal to an animal shelter (Scarlett *et al.*, 1999). In inner city areas, it is not unusual for cases of animal cruelty and abuse to be related to poverty and social problems. There are also specific inner city issues related in particular to dog fighting.

This activity is associated with other illegal activities, such as drugs or gangs, which are a marker for a general social malaise associated with poverty and all its consequences (Harding, 2014).

Excessive weight leading to obesity is becoming increasingly common in humans and their pets (German, 2006; Vis and Hylands, 2013). Such conditions have consequences for animal welfare (Courcier *et al.*, 2011; Becker *et al.*, 2012). Up to 40% of dogs in developed countries are overweight (Rohlf *et al.*, 2010). Several risk factors for obesity in companion animals have been identified (Kienzle *et al.*, 1998), including, among others, owner socioeconomic status (Robertson, 2003; McGreevy *et al.*, 2005) and change of lifestyle of the owner (Robertson, 2003), but also body condition score (BCS) of dog as judged by the owner and the BCS of the owner (Kienzle *et al.*, 1998).

In some cases, poor welfare can result from simple ignorance of animal biology. Uninformed owners may adopt mistaken beliefs about how their animals perceive the world around them, particularly if their attitude to pets is somewhat anthropomorphic. Anthropomorphism, defined by Serpell (2002) is the 'attribution of human mental states (thoughts, feelings, motivations and beliefs) to nonhuman animals'. If the owner's understanding of the animal's behaviour is inaccurate, the possibility arises that the animal may be treated in a way that the owner thinks is compatible with its welfare, but is actually detrimental (Bradshaw and Casey, 2007). Such anthropomorphic attitudes to pets is again a very common cause of animal obesity (above).

In addition to these more general examples of how human and animal welfare may be intimately interconnected and interdependent, it is also clear that concern for the welfare of animals is affected by the underlying attitudes towards animals of individuals, which may be a product of their own life-experience or underlying psychological character.

Concepts of attachment derive originally from human developmental psychology. The attachment between children and their caretakers, so-called primary attachment, is extremely important for a child's development. The type of attachment that is formed depends, among other factors, on the relationship a child has with its primary caregiver. Early attachment experiences are apparently carried forward, setting conditions for seeking, interpreting and reacting to later experiences. Thus the type of attachment formed (secure, preoccupied, dismissing, or fearful; Bartholomew, 1991) may influence the type of interpersonal relationships a person develops later in life. Fogany and Target (1997) found that without secure attachment a child may develop a personality disorder, characterised by enduring maladaptive patterns of behaviour, cognition and inner experience, exhibited across many contexts and deviating markedly from those accepted by the individual's culture. For example, if a primary caregiver is not available or is insufficiently available, a child might feel that he/she is not worth being looked after and cared for. If this feeling is elicited often enough, it will have a strong influence on the rest of that child's life and how he/she develops. The relationship between a person and an animal could also be considered a form of attachment in this sense (Gewirtz and Boyd, 1977). However, the human–animal bond cannot be explained in terms of primary attachment (Colby and Sherman, 2002), but in terms of other forms of attachment, such as adult attachment (Berman and Sperling, 1996).

Children develop an internal working model with attachment figures (Macfie et al., 2005). This internal working model makes it possible to predict the behaviour of the attachment figure and can later influence how children think of themselves. It is possible that children who grow up with animals develop an internal working model with a certain type of animal, for instance, a dog (Endenburg, 1995). If human caregivers are not always available, for whatever reason, children may form a secure attachment with a pet, which forms a protective factor during the child's development. Research has shown that children who grow up with a certain animal are more likely to own that type of animal in adulthood (Endenburg, 1991). Certain breeds of dogs – the so-called high-risk dogs – in a family can be an indication that domestic violence might occur. High-risk dogs are dogs who, without provocation, attack, bite or kill humans and/or other domestic animals (Barnes et al., 2006). Apart from being dangerous to humans and other animals, there is also a huge welfare problem with the dogs themselves. The way these animals are trained, injected with anabolic steroids, or killed when they lose a fight makes them also victims themselves (Harding, 2014). High-risk dogs are a part of a high-risk lifestyle. These dogs become a deviant possession much like a gun. Barnes et al. (2006) found in their study that owners of high-risk dogs had higher numbers of court convictions and had a wider range of deviant behaviours like aggression, problems with alcohol and drugs, crimes involving children and also domestic violence (Barnes et al., 2006). Owners of 'aggressive' dog breeds do indeed harbour more 'aggressive' personality traits than owners of 'non-aggressive' dog breeds (Ragatz et al., 2009; Wells and Hepper, 2012). The study of Schenk et al. (2012) further confirms that owning a high-risk dog seems to be an indicator of involvement in various antisocial acts.

It would appear that attitudes to animals may also reflect the human individual's attitudes to other humans. Animal abuse is a worldwide problem causing an incalculable degree of animal suffering (McMillan et al., 2015). Abuse is an intentional act that causes harm to an individual (McMillan et al., 2015). While international laws regarding animal welfare vary tremendously in their scope and detail and the consideration they may give to the responsibilities of individuals to promote positive welfare, almost all countries have legislation which makes it an offence deliberately to cause suffering (Chapter 7). Despite this, it is clear that abuse is widespread based on the (social) media messages regarding animal abuse which can be found rather regularly. Despite the vast amount of animal suffering involved, animal abuse has received scant attention as a subject of research (Sinclair *et al.*, 2006). Reasons for this are, among others, that veterinarians have difficulties recognising animal abuse, and when they do have a suspicion that it could be active abuse, they find it difficult to report it to the authorities in case their diagnosis is incorrect (Enders-Slegers and Janssens, 2009).

There is clear research, however, to demonstrate that animal abuse has a strong correlation with domestic violence (Lussier *et al.*, 2005), as well as with other criminal activities like human trafficking, drugs and dog fighting (Kalof and Taylor, 2007; Ragatz *et al.*, 2009). There is significant evidence to demonstrate that those who mistreat and abuse animals, show the same behaviour towards vulnerable people around them, such as children or the elderly (Ascione *et al.*, 2007; Ascione and Shapiro, 2009; Jordan and Lem, 2014). Acts of cruelty to animals may thus act as indicators of dysfunctions of human health and welfare. In a study by Baldry (2003), cases where domestic abuse was reported, which included physical violence and threats, were at least twice as likely to be related to animal abuse than non-violent cases. It is also often the case that those convicted of murder have a history of animal abuse (Garcia Pinillos *et al.*, 2016).

Domestic violence is violence performed in the family circle of the victim. The perpetrator can be a (ex)-partner, family members or a family friend. Forms of domestic violence are intimate partner violence, child abuse or abuse of elderly. In Ireland, 57% of women who were victims of domestic violence also witnessed cruelty or threats against their pets. The women remained longer in their abusive relationships because they feared for the well-being of their pets if they left. A majority (87%) would have left their partner sooner if their pets could also leave (Gallagher *et al.*, 2008). Domestic violence also has an influence on the development of children. As with child abuse, children's exposure to domestic violence can lead to short- and long-term outcomes of internalizing and externalizing behavioural problems during adolescence, including delinquency, status offences and perpetration of violence (Currie, 2006; Herrenkohl *et al.*, 2008). If violence between the parents is present, children are at an increased risk of being abused or neglected (McPhedran, 2009). Perhaps significantly, evidence shows that negative experiences within the family, or family dysfunction, represent a risk factor contributing to the development of childhood behavioural patterns which may lead them, in turn, to be predisposed to subsequent cruelty to animals (Shaw *et al.*, 2000; McCabe *et al.*, 2001). McPhedran (2009) stated that 'parental aggression towards a child, possibly resulting partially from a lack of parental empathy, may contribute to impairment in childhood empathy development, which can in turn contribute to the manifestation of cruelty to animals and other violent or "callous" behavior'.

The links between animal welfare, human well-being and the environment are complex. Colonius and Earley (2013) argue convincingly that: 'The separation between human, social, and animal welfare is an artificial compartmentalization. These disciplines rely on the same set of scientific measures and heavily depend on each other in an ecological context'. In this range of multidisciplinary areas, different professions and disciplines can work together to achieve common goals and improve both human and animal well-being. This recognition for the potential for cross-over between human and animal welfare and the factors affecting both has led to a concept of One Welfare (Garcia Pinillos *et al.*, 2016) as an extension of an existing recognition of the potential value of combining knowledge and expertise of veterinary and human medical knowledge within a broader concept of One Health adopted in 2007 by the American Medical Association (AMA) and the American Veterinary Medical Association (AVMA). A fully interdisciplinary approach to human and animal welfare in connection with the environment is necessary to further progress in welfare science. The multifaceted approach may also engage new fields that would provide additional perspectives, such as evolutionary geneticists, behavioural ecologists, psychologists and neuroscientists. Decisions should be made in an interdisciplinary frame with a focus on action and a mission of balancing and promoting human and animal welfare in connected ecosystems and societies (Mellor and Bayvel, 2008). The networking capacity of a One Welfare approach could empower the animal welfare field to more effectively address the connections between science and policy in various areas of human society (Coleman *et al.*, 1996; O'Riordan, 2004; Adam and Kriesi, 2007). A One Welfare approach will directly bolster connections and compassions between animal and human welfare (Colonius and Earley, 2013).

3

THE BIOLOGY OF WELFARE

The Brambell Committee's report on (farm) animal welfare in 1965 has had a dramatic influence on the assessment and management of animal welfare and casts a long shadow. While unquestionably a major advance at the time and heralding a significant change of attitude to the management and treatment of livestock animals, there have been many subsequent improvements in our understanding and scientific knowledge that require some re-examination of the Brambell principles and the concept of the Five Freedoms. Despite such advances, the Five Freedoms continue to dominate much of welfare practice and welfare thinking nearly 50 years after their original formulation.

In all fairness, the Brambell Committee's report never set out to be a 'welfare concept', but was developed specifically to establish minimum requirements to ensure the absence of negative welfare. The Committee formulated the idea that compromise to animal welfare is avoided if the animals are kept free from:

◊ hunger, thirst or inadequate food,

◊ thermal and physical discomfort,

◊ injuries or diseases,

◊ fear and chronic stress,

◊ and free to display normal, species-specific behavioural patterns.

There have been various reformulations of these essential principles (e.g. Webster, 1994), including an attempt by Mellor to substitute the concept of 'five domains of potential welfare compromise' (Mellor and Reid, 1994). The five domains are defined in terms of nutrition, environment, health, behaviour and mental state).[2] But these different incarnations change the basis of the construct little and, as an instant index, the Five Freedoms remain widely used today as a guideline for welfare assessment protocols, with the actual state of welfare of an animal being characterised as unimpaired if it complies with the Five Freedoms (e.g. Rutherford, 2002; Veissier and Boissy, 2007; Knierim and Winckler, 2009; Mendl *et al.*, 2010). In fact, the Five Freedoms form the basis for the so-called Welfare Quality Project, which currently forms the backbone of European animal welfare guidelines (http://ec.europa.eu/food/animal/welfare/sum_proceed_wq_conf_en.pdf).

We should emphasise, however, that these freedoms were primarily derived in relation to the welfare of farm animals and may only have restricted utility when applied to animals whose environment is less rigorously controlled by human intervention. Except in regard to the fifth freedom, the animal is conceived as undergoing its personal life conditions somewhat passively. This was perhaps legitimate in that Brambell's freedoms were originally developed primarily for application to domestic animals or animals whose environment was otherwise largely controlled by human management. But if these freedoms are to be more widely applied to the assessment of welfare of less closely managed or truly free-ranging animals, then they are somewhat over-restrictive (see reviews by Ohl and Putman, 2013a, 2013b, 2014b).

A biological basis of animal welfare

More recent analyses of animal welfare acknowledge the fact that at any one point in time, an individual animal's welfare status lies on the continuum between negative/bad welfare and positive/good welfare (i.e. well-being) (e.g. Dawkins, 2008; Yeates and Main, 2008). Despite this agreement, the majority of current approaches to safeguarding animal welfare still tends to focus on the exclusion of factors that are understood as compromising welfare by being 'negative' for the animal, such as being ill, wounded or stressed. The majority of animal welfare scientists agree, however, that this persistent emphasis on the avoidance of negative states is somewhat minimalist – and over-restrictive in that it ignores the active promotion of positive welfare or well-being and masks consideration of the fact that in assessing and delivering welfare, we must pay equal attention to those factors that may help promote positive welfare as much as avoidance of negative welfare states.

---

2 although this latter idea has not been widely adopted, other than by Mellor himself in subsequent publications: (Mellor and Stafford, 2001; Mellor, 2004; Mellor et al., 2009).

Even in 1998, Fraser and Duncan noted that public discussion on animal welfare suggested:

> (i) that minimising suffering (strong, negative affective states such as severe hunger, pain, or fear) is of primary concern; (ii) that allowing animals to experience normal pleasures of life is considered relevant to welfare [*although*] of lower priority than prevention of suffering; and (iii) that factors producing no affective response in the animal are seen as either not relevant or of reduced relevance to animal welfare.

Accordingly, throughout this book we emphasise that welfare must be considered as more than simply the avoidance of negative states, but that any welfare concept should extend to embrace promotion of positive status (Fraser and Duncan, 1998; Boissy *et al.*, 2007; Dawkins, 2008; Yeates and Main, 2008; Mellor, 2012; Ohl and van der Staay, 2012). And, while most legislative provision does indeed remain directed towards the avoidance of management practice which might result in suffering, there has more recently been a greater emphasis given to the view that good animal welfare requires the presence of positive affective states, as well as the absence of negative ones (Duncan and Dawkins, 1983; Duncan, 2005; Broom, 2010). Recognising this, it would seem that society increasingly expects that management action will seek proactively to enhance positive welfare status rather than simply avoid negative welfare.

More fundamentally, a view of welfare which is dominated by an emphasis on the avoidance of negative states neglects the fact that, except in the specific instances where natural selection processes have been largely countermanded by deliberate selection by humans, animals have evolved, optimising the ability to interact with and adapt to (changing conditions within) their environment, and that exposure to environmental challenge and short periods of 'negative welfare' may therefore be inevitable if these are understood as triggers to release from the animal's repertoire the appropriate behavioural or physiological response to adapt to those challenges (see, for example, Barnett and Hemsworth, 1990, among others).

Welfare as related to an animal's adaptive capacity

More recently a number of authors have advocated a more dynamic view of welfare which recognises that both wild and domestic animals have adaptive responses that enable them in normal conditions to respond appropriately to address some environmental or physiological challenge and to restore a more positive welfare state (except perhaps in those cases where artificial selection may have resulted in the loss of some responses from domestic stock). This idea that animals have generally evolved adaptations to their environment, optimising the ability to adapt to changes within that environment through the expression of a variety of physiological and/or behavioural responses, was first applied within a welfare context some three decades ago (see, for example, Carpenter, 1980; Dantzer and Mormede, 1983; Broom, 1988; Barnett and Hemsworth, 1990) and has more recently been championed by, for example, Duncan (1993), Fraser *et al.* (1997), Fraser and Duncan (1998), Korte *et al.*, (2007) and Ohl and van der Staay (2012), among others.

Indeed, it seems remarkable that it was as long ago as 1980 that Carpenter wrote: 'The welfare of managed animals relates to the degree to which they can adapt without suffering to the environments designated by man. So long as a species remains within the limits of the environmental range to which it can adapt, its well-being is assured'. Although nowadays we might not consider such adaptive capacity at the level of the 'species' but rather at the level of the actual adaptive capacity of each individual, the basic principle is unchanged: in such a concept, the animal's welfare is not at risk as long as it is able to meet environmental challenges, i.e. 'when the regulatory range of allostatic mechanisms matches the environmental demands' (Korte *et al.*, 2007).

Despite the fact that this idea was first mooted so many years ago, it has taken some time for the concept to be more generally adopted, and it is still by no means universally reflected in the literature. However, many now do advocate this more dynamic view of welfare, with the implication that a welfare problem arises only when an animal or a group of animals have insufficient opportunity (freedom) to respond appropriately to a potential welfare 'challenge' through adaptation by changes in its own behaviour (e.g. Broom, 2006; Korte *et al.*, 2007; Ohl and van der Staay, 2012).

On this basis, we may suggest that a positive, or at least non-negative, welfare state would be safeguarded when the animal has adequate freedom to react to the demands of the prevailing environmental circumstances, resulting in a state that the animal itself perceives as positive. In such a view, assessment of welfare should therefore focus not so much on the challenges which any animal may face at a given moment, but on whether or not the animal has the freedom and capacity to react appropriately (i.e. adaptively) to both positive and potentially harmful (negative) stimuli (Ohl and van der Staay, 2012). From such a perspective, welfare should not be considered as a static, or rather passive status, but should be considered in relation to the adaptive capacity of the animal or animals concerned.

As noted by Broom (2006), 'A key point concerning the concept of individual adaptation in relation to welfare is that welfare may be good or poor while adaptation is occurring'. By extension, this implies that welfare should not be considered as an instantaneous construct to be assessed at some moment in time. An adaptive response may take some finite period of time; crucially, therefore, our assessment of welfare does not simply consider the status of any individual at a given moment in time, but needs to be integrated over the longer time periods required to execute such change.

As noted, this more functional, adaptive view of welfare and its assessment is NOT necessarily embraced by all commentators and, in the interests of balance within this review, it is important to emphasise that some authors still adhere to the more traditional views of welfare enshrined in various modifications of the Brambellian freedoms. However, these more adaptive approaches are gaining ground and, in truth, do represent a more biological basis for considerations of welfare status.

Fitness and welfare

This idea of relating the concept of welfare to the adaptive capacities of an individual has previously resulted in considerations on the relation between welfare and (Darwinian) fitness – and in fact the concepts of fitness and welfare have even been equated by some authors (see review by Appleby and Sandøe, 2002; see also Jordan, 2005; McGreevy and Bennett, 2010; Brosnan, 2011).

Jordan (2005) stated that 'because it is impossible to make a detailed physiological study in free-living animals, reliance must be placed on behaviour and Darwinian fitness, which have been shown to correlate to welfare'. Further, Fraser and Broom (1990) considered that 'the Darwinian fitness of an animal may be used to evaluate whether an animal is coping'. Hofer and East (1998) suggested that a Darwinian approach is essential for any theory of stress in biological conservation and further believe that using a reduction in Darwinian fitness to define stress has a number of advantages: 'the approach has universality, does not require specific mechanisms that are restrictive, and recognises that animals employ a variety of mechanisms to cope'.

Others, however, emphasise that individual fitness and welfare are not the same (Dawkins, 1998; Webster, 1994). Dawkins, as an example, explains that seeking shelter in mice is of fitness value in nature but not under captive conditions, as the latter conditions would not include any predator risk.

Nevertheless, she argues, welfare in mice would be improved under captive conditions by providing them with shelter. Such a conclusion, implying that an individuals' welfare status might be increased by provision of a resource which no longer contributes anything to fitness, makes it clear that fitness and welfare cannot be equated.

There may, moreover, be other examples arguing against an equation of fitness and welfare. As discussed for example by Mendl and Deag (1995) a dominant position within a social group is likely to maximise fitness in terms of reproductive success, but being of high dominance often means being continually challenged for that key position, thus increasing, for example, the risk of injury or reducing the time to find sufficient food. Moreover, dominant individuals may be chronically stressed (Sapolsky, 1993; Ohl and Fuchs, 1999) which may result in significant impairment of general welfare (Barnett and Hemsworth, 1990). Even if both fitness and welfare are *sometimes* favoured by the same factors (as in the example of shelter seeking in mice), both are also affected by other – very different – factors and, by the same token, may not infrequently conflict (Barnett and Hemsworth, 1990; Mendl and Deag, 1995).

We must in fact recognise (after Korte *et al.*, 2009) that natural selection exerts genetic benefits by maximising reproductive success of the adapted organisms, even at the expense of individual happiness, health and longevity. In effect, if we presume welfare to mean positive or negative 'well-being' (e.g. Broom, 2007), then that welfare status may be compromised without affecting the ability to leave offspring (unhappy animals can still reproduce successfully). By converse, even if fitness is compromised in some way and, as a consequence, the animal's ability to leave progeny is reduced, this fitness issue is not *necessarily* of welfare concern.

From all this (and see also the review by Mendl and Deag, 1995), it seems clear to us that welfare is not identical to fitness. But following the concept that animal welfare might be defined by the animal's freedom to adapt to changing environmental circumstances, perhaps welfare might depend on the animals' ability to carry out behavioural routines that, under natural conditions, may simultaneously enhance fitness.

Welfare as the animal's own perception of its status

It is clear that emotions play an important role in the performance of adaptive behaviours – indeed there is a considerable literature to suggest that much of the function of emotion or emotional status may be to provide a convenient proximate surrogate to reinforce behaviours which are (or were) in some way adaptive, such that performance of these appropriate behaviours is in some sense rewarding. (The arguments for the evolution of feelings as part of animal functioning are rehearsed by Nesse and Ellsworth, 2009; Panksepp, 2011; Webster, 2011). If the performance of fitness-related behaviours is thus in some sense rewarding (and lack of opportunity to perform such behaviours, by converse, in some way frustrating or actively damaging), then the welfare of a mouse in captivity will indeed be enhanced by providing it with shelter, even though its fitness will not be enhanced under the actual management conditions.

Most modern commentators would now acknowledge that, to some significant degree, any animal's status must be that perceived and judged by that animal itself (Duncan, 1993; Fraser and Duncan, 1998; Broom, 2006; Taylor and Mills, 2007; Nordenfelt, 2011; Webster, 2011). It is perhaps self-evident that if some animal is perceived by the observer to be in a negative welfare status, but has opportunities (correct behavioural repertoire, appropriate environmental conditions) to improve its status, yet fails to take that action, then it may simply perceive its own status as satisfactory. In effect, the 'decision' by any individual animal to accept its current status or to engage in behaviour designed to bring about some change of status must in part be determined by an assessment of physiological condition (hunger, thirst, etc.) but also by an assessment of a sense of 'well-being'.

It is clear that emotions play an important role in this assessment and in the performance of adaptive behaviours. As suggested earlier, there is a growing literature to suggest that much of the function of emotion or emotional status may indeed be to provide a convenient proximate surrogate to reinforce behaviours which are (or were) in some way adaptive, to make performance of these appropriate behaviours in some sense pleasurable or rewarding and thus promote their expression in appropriate circumstances (Cabanac, 1971, 1979; Broom, 1991, 1998; Mendl and Deag, 1995; Dawkins, 1998; Panksepp, 1998, 2011; Lahti, 2003; Webster, 2011).

As nicely summarised by Nesse and Ellsworth (2009), 'Emotions are modes of functioning, shaped by natural selection, that coordinate physiological, cognitive, motivational, behavioral, and subjective responses in patterns that increase the ability to meet the adaptive challenges of situations that have recurred over evolutionary time'. In other words, emotions would appear to play a pivotal role in decoding and valuing reward and punishment, in perceiving the individual's own emotional state and finally in regulating the execution of its behaviour. What is significant in this context, however, is that there is clear variation between individuals in sensitivity/responsiveness of central nervous circuits processing emotions and that different individuals may 'perceive' the consequences for themselves of one and the same environmental challenge in very different ways. We will consider the implications of this further in Chapter 4.

4

INDIVIDUAL VARIATION IN SELF-ASSESSMENT AND IN COPING STRATEGIES

Variation in self-assessment

As we have just noted, it would appear that emotions play a pivotal role in decoding and valuing reward and punishment, in perceiving the individual's own affective state and, finally, in regulating the execution of its behaviour (Ohl and Putman, 2013c). What is significant in this context, however, is that there is clear variation between individuals in sensitivity/responsiveness of central nervous circuits processing emotions and thus that different individuals may 'perceive' the consequences for themselves of one and the same environmental challenge in very different ways.

Variation in coping strategies

There may also be quite substantial variation in the way different individuals may respond to the same stressor and the strategies they may use to cope with environmental or social challenge. The adaptive repertoire of different individuals will itself show significant variation because both physiological and behavioural responses will have, in part, a genetic basis (the basic repertoire may thus itself vary between individuals) but may also be developed differently through differences in ontogenetic experience. Furthermore, adaptive responses may be modified in subsequent life, or novel responses developed *de novo* through learning; and unquestionably different individuals will be exposed to significant differences in experience and learning opportunities through life.

Finally, it is clear that whatever underlying differences there may be in the adaptive repertoire available to different individuals, there are also marked differences in relation to the whole 'style' of adaptive strategy adopted. Observation of behaviour of individual animals over extended periods of time have revealed intriguing variation in behaviour within single populations. Increasing evidence suggests that there are marked differences between individuals in their behaviour and, importantly, that these differences in behavioural 'style' are consistent across time and across a range of contexts (Sih *et al.*, 2004a, 2004b; Réale *et al.*, 2007). This intraspecific variation was previously treated as statistical noise, but behavioural ecologists and animal welfare scientists now realise that these 'personality' types may be adaptive and useful for management (Mendl and Deag, 1995; Smith and Blumstein, 2008; Blumstein, 2010). In some cases, recognition of such differences in 'personality' (Gosling, 2001) has been somewhat subjective and qualitative, with risk perhaps of projection onto the animals observed, of some anthropomorphic judgement of the observer, but there are a number of more formal analyses of this sort of variation in response to a given situation with examples presented for cats (Feaver *et al.*, 1986), hyaenas (Gosling, 1998), bears (Fagen and Fagen, 1996), fallow deer (Bergvall *et al.*, 2011) and great apes such as chimpanzees or orang-utans (King, 1999; Weiss *et al.*, 2006). In addition, it has even been suggested that there may be some positive correlation between such different behavioural styles (more formally behavioural 'syndromes') and variation in cognitive style or ability (Sih and del Guidice, 2012).

One critical subset of this more general variation in behavioural style is in the way in which individuals may respond to external challenges or stresses. Dingemanse *et al.* (2009) make a distinction between the more general variation in behavioural style (animal 'personality' overall) and 'coping style': the suite of behavioural and physiological responses of an individual that characterise its reactions to a range of stressful situations. As illustration here, extensive behavioural and physiological analyses show that there exist clear and distinct strategies, or styles, in the way in which male mice may cope with environmental challenges. In a series of studies, Koolhaas (2008) and Koolhaas *et al.* (2008, 2010) have presented experimental evidence for significant individual variation in coping style which is closely related to predictable variation in behaviour and underlying physiology: aggressive male mice tend to cope actively with their environment whereas non-aggressive males seem more easily to accept the situation as it is. 'Reactive' copers respond to changes in their external environment, whereas 'proactive' copers respond less or not at all.

Such variation is perhaps not unexpected because, after all, individual variation is accepted as a necessary prerequisite for evolution. Notably, in nature, mouse populations are known to go through phases of growth and decline and it seems that while the more aggressive, actively coping males dominate the group during the phase of colony-growth, the non-aggressive phenotype is more successful in establishing a new colony than the highly aggressive phenotype (Korte *et al.*, 2005). Given that non-aggressive males indeed 'find it easier to accept the situation as it is' (Koolhaas *et al.*, 2010) we may conclude that their adaptive capacity is higher than that of individuals of the aggressive phenotype (Coppens *et al.*, 2010). Further, aggressive and non-aggressive male mice have been demonstrated to differ in terms of emotional processing and it seems reasonable to conclude that non-aggressive, passively coping males perceive their subordinate situation within any social hierarchy differently from their aggressive counterparts. Similar evidence of the existence for distinct coping styles within wild populations has been provided in, for example, bighorn sheep (Réale *et al.*, 2000) or in alpine marmots (*Marmota marmota*; Ferrari *et al.* (2013).

Whatever the detail of these examples, the significance is a clear recognition of substantial individual variation in coping strategy and the way in which individuals respond to or deal with the same environmental challenge. That variation in coping strategy must itself imply clear difference in the impact of one and the same stressor on the welfare of those different individuals.

Optimising or satisficing?

Clearly, there is growing evidence that individual animals may show significant variability in the way they perceive, or respond, to the same challenges to welfare status. In addition, we should not necessarily presume that all individuals seek to maximise welfare status at any given instant of time.

At this point, it is of relevance to consider the concept of 'satisficing', which was developed by Simon (1955) as a psychological concept and subsequently applied by ecologists to explain foraging behaviour in animals (Myers, 1983; Ward, 1992, 1993). Satisficing is an alternative to optimisation for cases where there are potentially many possible alternative options which cannot effectively be fully evaluated. A decision maker who gives up the idea of obtaining an optimal solution but obtains the optimum he can compute under given time or resource constraints is said to satisfice (Krippendorf, 1986).

In this approach one sets lower bounds for whatever various objectives that, if attained, will be 'good enough' and then seeks a solution that will exceed these bounds. The satisficer's philosophy is that in real-world problems there are often too many uncertainties and conflicts in values for there to be a realistic probability of obtaining a true optimisation and that it is far more sensible to set out to do 'well enough' (but better than has been done previously).

Satisficing thus should not be understood as being satisfied with the absence of a 'negative state'; in contrast, this concept explains behavioural strategies that allow for reaching an optimum state by accounting for prevailing environmental conditions. As we have outlined earlier, a variety of scenarios seem to exist, at least for socially living animals, where the behavioural strategy of particular group members is more likely to be explained by a strategy of optimising or satisficing its own state of welfare than by assuming that a universal maximum state of welfare is to be reached.

The logical extension of such argument is now clear: if, as seems apparent, there exists significant variability in how different individual animals may assess their own welfare status; if, in addition, a group of animals may consist of individuals some of whom seek to optimise welfare status while others (at that moment in time) are content to satisfice, then we, as external observers, may expect to observe within any group of animals, a series of group members with different 'absolute' welfare status (assessed by an external observer against some fixed set of criteria) which nonetheless perceive their own welfare state as being optimal (or at least sufficient not to require action to alter that status). In such a case then, the (objectively determined) welfare status of all members of that group may appear to vary over a considerable range, yet all members perceive their own welfare state as optimal – or at least satisfactory.

Such a conclusion also makes clear that purely objective functional scales for measuring the welfare status of individual animals can have little validity in that, even under identical conditions, the actual welfare status of different individuals may vary widely. Further, it emphasises that when assessing the welfare status of animals in groups or populations we must expect high variation in apparent welfare and in attempting to safeguard satisfactory welfare we must insert into protocols some minimum threshold value below which no individual should be allowed to fall, instead of (or in addition to) simply determining some average welfare status to be achieved.

5

Welfare at the Group Level

Many animals live in social groups – the other members of that social group, and interactions with those other group members, are as much a part of the individual's environment as its immediate physical surroundings. It is widely accepted that the behaviour of other individuals may influence the behaviour and welfare of others within their social group. Likewise, because group members have some interdependence, the welfare of one individual may be enhanced or compromised by the welfare status or management 'applied' to others within its social group.

At its simplest: welfare of a juvenile may be compromised by death or injury of its mother, or by deliberate forced weaning; welfare of even adult members of a group may be compromised by removal or loss of a 'lead individual' whose experience guides the group in routine foraging or social interaction. Death of the matriarch in a group of elephants may have serious implications for the welfare of younger and less experienced females in that social group, whose welfare had been previously dependent on her knowledge and expertise to lead them to new food sources or to water in periods of drought. Thus welfare of any individual may be affected by the welfare status of others within its social group (or indeed the wider population of which it is a part).

Various lines of analysis suggest that the 'fitness' of any individual organism cannot be effectively assessed at the level of that individual, but that, in practice, true evolutionary or genetic fitness needs to consider the 'inclusive fitness' of the individual and its close relatives (Hamilton, 1964a, 1964b), which are often members of the same population or social group. The actual fitness of the group may therefore be of greater significance than the apparent fitness of the individual. Within a social group or population of animals, a series of behaviours may be observed which appear to be 'altruistic';

that is to say that such behaviours appear to enhance the fitness of other individuals within the group at the expense of a possible loss of fitness of the individual performing that behaviour (see, for example, a review of such behaviours offered by Krebs and Davies (1993) or a more recent assessment of the 'fitness value' of such behaviours by Clutton-Brock, 2002).

Such 'altruistic acts' have variously been explained biologically on the basis of an expectation of reciprocal behaviour (if the trait is dominant within the group/population) –'reciprocal altruism' as defined by Trivers (1971) – or that the behaviour in some indirect way does enhance individual fitness.[3] Others hypothesise that 'the investment in altruism is like an investment in advertisement. It should depend on its effect on the target audience and on the potential effect of this audience on the fitness of the advertiser' (Zahavi, 1995; Lotem *et al.*, 2003).

But the simplest explanation for such apparent altruism in many instances is in recognition of a concept of inclusive fitness whereby, by increasing the fitness of close relatives or directly assisting the reproductive efforts of close relatives, an individual increases the copies of its own genes passed on to successive generations, because those 'helped' share significant numbers of the same genes as the 'helper'. Indeed, in the extreme, increasing the fitness of close relatives may, on occasion, be a more economical way of passing on copies of one's own genes than engaging directly in reproduction for oneself, as long as the coefficient of relatedness is high (Hamilton, 1964a, 1964b).[4]

In the same context it is known that a variety of behaviours exist within social groups that appear to be directed towards increasing the competitive fitness of the group rather than simply that of the individual itself (West *et al.*, 2007; Crowley and Baik, 2010) – although clearly such behaviours may also increase the fitness of the individual through its continued membership of a more successful group. One such example of the acceptance of risk apparently on behalf of the group is the intriguing case of voluntary foraging by a queen in the ant *Acromyrmex versicolor* (Rissing *et al.*, 1989; Seger, 1989).[5]

If we now re-evaluate a potential relation between fitness and welfare, we must recognise that the inclusive fitness of the individual may be achieved at a point where the welfare of all member individuals might not

---

3 For example, subadults assisting adults in some form of cooperative breeding may be learning and thus increasing the probability of success of their own first breeding attempt or may be increasing the likelihood of inheriting the breeding territory or mating rights of those individuals who have been helped.

4 However, see a re-assessment of the importance of relatedness in explaining apparent altruism offered by Clutton-Brock (2002).

5 See also Crowley and Baik (2010) for fuller treatment of this so-called 'public goods game'.

be individually at maximum. As an example, we might consider again the situation of individuals living within a social hierarchy. To achieve a stable social hierarchy there have to be, within that hierarchy, dominant as well as more subordinate individuals. Irrespective of the question whether being dominant may result in more positive welfare or not (see, for example, explorations by Mendl and Deag, 1995), life conditions of group members will differ in practice depending on their social rank and it seems reasonable to assume that not all group members will be able to maximise their welfare state within such a social hierarchy.

These considerations result in the question of whether there may be a place for an 'inclusive welfare' concept, somewhat akin to the concept of inclusive fitness (Hamilton, 1964a, 1964b) which embraces the welfare of close relatives as well as that of the individual. Such a concept indeed may be of potential use for explaining behavioural phenomena that are in conflict with more conventional welfare concepts, such as the performance of behaviour that seems to improve the welfare of group members rather than that of the behaving individual(s). Evaluating the possible nature of such an inclusive individual welfare inevitably leads to the question of whether there may be a concept of (inclusive) *group* welfare – that the welfare interests of the individual may include the welfare interests of relatives. If we may assume that welfare phenotypes exist and are sensitive to evolutionary mechanisms at the level of the individual, it may be useful to further evaluate whether the existence of some concept of 'inclusive welfare' could actually explain the occurrence of these potential inter-individual differences in welfare profiles.

We may postulate that from a welfare point of view it might well be biologically appropriate to invest into the welfare of other members of one's social group if such behaviour strengthens the functioning of the group as a whole (and thus increases the welfare of the individual performing such group-enhancing behaviour, as a member of that group). Performance of 'other-regarding' behaviours (*sensu* Lahti, 2003) may in any event improve the welfare status of the individual performing such behaviours if, in some instances, internal reward mechanisms may be triggered not only by selfish behaviours but by behaviours directed to others within the group. In this context, we may also note that social affiliation in general has been shown to activate reward mechanisms in the brain (Kikusi *et al.*, 2006), confirming that social investment may facilitate a positive perception of its own internal state in the investing individual. Investment behaviour might thus be explained within such a concept of inclusive welfare in the sense that the individual's perception of its own internal state may be optimised by investing in the welfare of other individuals as the result of the self-rewarding characteristics of such behaviour.

A recent review of allogrooming in primates (Crofoot *et al.*, 2011) revealed that 'extensive groomers who tend to direct their efforts at other extensive groomers also spend time in close proximity to many other individuals'. These results indicate the important role that prosociality plays in shaping female social relationships. Significantly, Crofoot *et al.* (2011) also reported that 'females who receive the least aggression, and thus pay low costs for group living, are most likely to participate in group defence'. From such results we may conclude that, within a social hierarchy, investments into the welfare of other group members may take place in different form: while some group members may invest by engaging in socio-positive behaviour, others may invest by defending the group.

This leads us to conclude that for such social species, we do indeed need some concept of welfare that recognises the fact that the welfare of a focal animal depends to some degree on the welfare of its social partners and argue that, at least among social species, individual welfare should be re-evaluated as being related to the individual's functioning within a social group and, moreover, to the functioning of such a social group. We will argue later that, at least among social groups, animals with different 'welfare strategies' may optimise their perceived internal state and, consequently, their own welfare status, in performance of a varying balance of both self-regarding and other-regarding behaviours (see also Crowley and Baik, 2010).

To pursue this argument, we now need to explore whether such a concept holds true for individuals living in social groups more generally and, further, how to assess the welfare state of such a social group. One of the first attempts to measure welfare at the group level was offered by Kirkwood *et al.* (1994) in populations of wild mammals and birds. These authors argued that an

> 'Assessment of the scale and severity of harm to welfare requires consideration of several factors. We propose that at the simplest level these are: 1. The number of animals affected. 2. The cause and nature of the harm. 3. The duration of the harm. 4. The capacity of the animal to suffer'.

This approach implicitly presupposed that a given cause and type of harm with a given duration will result in an identical effect on welfare in all individuals of one group, as long as all individuals have the same capacity to suffer (Kirkwood *et al.* (1994) and Mathews (2010) presume that they do).

Consequently, this scenario suggests a linear correlation between group size/ animal number and the scale of harm caused to (group) overall welfare by any given challenge. Put in another way, this implies that the welfare status of a group could be considered as being represented simply as the sum of the individual welfare of its members.

This calculation in practice assumes homogeneity of all members of a group in terms of state, sensitivity and perception of welfare at any point in time – a 'universal' individual welfare. As argued earlier, individual differences in adaptional repertoire and differences in coping strategy point towards the idea that welfare of each individual should be related to the adaptive capacity of that individual, making it more likely that distinct external conditions will affect the welfare of the members of any one group to a *different* extent. While this refines the concept of Kirkwood *et al.*, it still leads to an idea of group welfare as the sum of some external assessment of the (objective) welfare states of individual members. We would suggest, however, that this focus on an objective assessment of welfare of each individual against some fixed and external scale of values overlooks the fact that the actual welfare status relevant to each individual is, in fact, more related to the individual's own perception of its welfare state (see Chapter 3). This implies not only , as above, that the welfare status of all individuals in a given set of conditions may not all be equal in absolute terms but that, in addition, the actual effect (perception) of any compromise in welfare status may not be the same for all individuals because they may vary phenotypically in the extent to which they respond to a given (positive or negative) influence.

Group benefit – individual sacrifice?

Despite such recognition that the actual welfare status of each individual – and, more significantly, the welfare status it experiences – may show significant variation between individual members of any group, the question follows of whether optimising (inclusive) group welfare demands (voluntary) compromising of the individual welfare of some of the group members. If we assume the existence of a 'universal' individual welfare that is true for all individuals, this would indeed have to be postulated. If, however, we accept that there is a considerable variation in welfare profiles of different individuals even under identical conditions, optimum group welfare could be obtained by a combination of group members with different 'absolute' welfare status (assessed by an external observer against some fixed set of criteria), yet which each perceive their own welfare state as being optimal – or at least satisfactory.

If we accept (as in Chapter 1) that the individual's perception of its own state as being positive, defines positive welfare and also that every individual seeks to reach and maintain a state of positive welfare (Duncan, 1993, 1996; Dawkins, 1998), we may expect that welfare-related behaviour is likely to be characterised primarily by behaviours that are related to the improvement of the individual's own welfare; in other words, self-rewarding behaviour. However, if we now postulate that welfare-related activity may also include performance of behaviours directed towards improving the welfare of others within the same social group, it seems to us likely that individuals within any group are likely to show significant variation in the extent to which internal reward mechanisms are triggered by self-regarding or other-regarding behaviour. Different behavioural phenotypes, at least within social groups, may therefore optimise their perceived internal state and, thus, their own welfare state by behaving differently as social partners (e.g. Crofoot *et al.*, 2011). Welfare phenotypes may therefore vary along a continuum between self-regarding and other-regarding behaviour at the executive level, which would be in line with considerations of biological altruism and echoes observations of Mendl and Deag (1995) on the possibility, within groups of animals, of some frequency-dependent stability of alternative coping strategies.

We suggest that, in the same way that the welfare status of an individual animal is best assessed by observations of its behaviour and adaptive responses (Korte *et al.*, 2005; Koolhaas *et al.*, 2010; Ohl and van der Staay, 2012), so too, this emergent property of 'group welfare' (that which is over and above the sum of the individual welfare states of its individual members) is best measured by the extent to which the group as a whole may show adaptiveness and respond as a group to adverse or suboptimal environmental conditions (see Box feature at end of chapter).

Conclusion

When exploring the various mechanisms ensuring adequate or optimal animal welfare, it is necessary, at least among more social species, to re-evaluate individual welfare as being related to the functioning of a social group (Mendl and Deag, 1995). We propose here a concept of inclusive welfare, taking into account that, at least in socially living animals, a variety of situations exist where individuals invest into the welfare of other individuals (often closely related) instead of necessarily maximising their own state of welfare. We further argue that such inclusive welfare might have been developed through a coupling of investment behaviour with internal reward mechanisms, facilitating the performance of distinct investment behaviours that are likely to result in a long-term pay off for the investing individual. Since the stimulation of internal reward mechanisms represents strong internal incentive, such investment behaviour results in a positive perception of the individual's own internal state.

To explain a potential evolutionary advantage of inclusive individual welfare, some distinct concept of group welfare needs to be considered. We suggest that welfare phenotypes may exist that vary along a continuum between self-regarding and other-regarding behaviour at the executive level. We hypothesise that optimum group welfare might be obtained by a combination of group members that differ in their perception of absolute welfare states and thus perceive their own welfare as being optimal under different individual circumstances. Group welfare may therefore represent an Evolutionary Stable Strategy (ESS; Maynard Smith, 1982; Parker, 1984) in that such an ESS will be reached if the individual welfare of all its members is optimised or at least satisficed (*sensu* Krippendorf, 1986), and variance in individual welfare perception is in accordance with a stable balance of self-regarding and other-regarding individuals under prevailing environmental conditions.

At the individual level, we assume that positive welfare is defined by the animal's freedom to adapt to environmental conditions up to a level that it perceives as positive. But we recognise that individuals may show significant variation in their perception of a given status and their 'decision' about how to respond to that perceived status. We may therefore expect that even under identical environmental conditions, different individuals within a group or population may perceive or experience their welfare status differently. Different individuals within a group or population may thus be 'satisfied' with different levels of what an external observer would consider better or worse states. When assessing the welfare status of animals in groups or populations, we must therefore expect high variation in apparent welfare.

Management implications

As a result of this variability between individuals, all of whom may perceive their own welfare status as optimal, or at least satisfactory, welfare of social groups cannot adequately be assessed by assuming that the impact on all individual group members will be identical. Instead, it is important to consider the welfare impact of any given challenge (or set of challenges) upon individual members of a social group which are not seen as identical clones but may vary considerably in their 'welfare phenotype'. It is important also to consider the possible impact on the group as a whole in terms of its capacity as a group to adapt to prevailing environmental conditions because of a given phenotypic distribution.

The welfare status of a group may thus be defined by the freedom to adapt adequately to prevailing environmental circumstance, both as individuals and as a group. Group welfare may be optimised, while the (objectively determined) welfare states of its individual members may vary over a considerable range (yet with all members perceiving their own welfare state as being optimal). By corollary, to ensure positive welfare of such a group, we cannot necessarily expect to optimise welfare status of all individuals. But that variability in itself implies that in attempting to manage the welfare status of a group or larger population, we may need to set an additional constraint which ensures that the status of no individual falls below some critical minimum threshold.

ADAPTIVE CAPACITY AT THE LEVEL OF THE GROUP

What exactly do we mean by adaptive capacity of a group? Is it possible that there is some emergent property at group level which is greater than the sum of the adaptations of its members (but nonetheless derives in some way from the abilities of those member individuals)? Might groups have different adaptive capacities based on the capacities of their members? Such a construct does not necessarily have to invoke ideas of 'group selection' because the capacity of the group remains a function of the adaptive abilities of its members – but how could it work?

A surprising number of animals make, or appear to make, group decisions. Herbert Prins, for example, has described how members of herds of female African buffalo (*Synceros caffer*) may 'vote' on where to move within their range to feed (Prins, 1996). As preferential grazers within a generally wooded environment, buffalo tend to feed in grassy clearings scattered at intervals within a wider habitat matrix of woodland and scrub. In general, these grassy areas are comparatively small and because females move in quite large matriarchal groups, quality forage is relatively quickly exhausted and the herd has to move on to another clearing. The decision of which patch to visit next is a complex function of the overall quality of each patch, the productivity – and the time since the group last visited any given patch to feed (thus depleting the availability of quality forage).

Prins observed each group and came to the conclusion that the decision of which patch within their overall range the group should visit next was made communally, with a rather subtle system where each individual cast her own vote. Essentially, as ruminants, buffalo feed for a few hours and then lie down to rest and ruminate, processing the forage taken in during the previous bout of grazing. At intervals, each animal rises to stretch, perhaps to defaecate, before lying down again to continue cudding. Gradually, over time, Prins observed that as each animal rose and lay down again, the whole group assumed a common orientation, tending to face more or less in the same direction. Eventually, they would all get up and move off to a new grass patch in the direction in which by now they were all facing (or, technically, because they were not all facing *exactly* the same direction, they moved off in the direction indicated by the mean vector calculated as the product of the vectors of all herd members combined). Prins was convinced that a group decision had been taken as to where to travel for the next feeding bout. In some groups, the orientation of all individual members was actually tightly clustered around that mean vector; in other groups, individual members faced a much wider range of directions, so that the mean vector of travel was more difficult to calculate and agreement on the 'decision' was less clear.

There is no evidence to test whether those groups which made more consistent decisions 'chose' better than those where agreement was less clear, but the fact that group decisions seemed to be made at all obviously offers a potential that the decision-making ability of groups (or the group's adaptive capacity) might well differ between groups, with possibly a real potential effect on both fitness and welfare at the group level.

Image credit: Haplochromis

6

WELFARE STATUS VERSUS WELFARE ISSUE:
THE IMPORTANCE OF ETHICAL CONSIDERATIONS

At the end of such reflection on the biological basis of welfare and the factors affecting the welfare status of individuals and groups, we may perhaps attempt some new definitions to replace those rather vague generalisations in Chapter 1. Based on the considerations explored earlier, we would propose the following definitions.

Adaptive capacity

The adaptive capacity describes the set of physical and mental abilities with which an animal is able to respond and 'adapt' to its environmental situation and any challenges it may encounter. Many features of this adaptive capacity have been acquired by a species through evolution; others may be developed by individual animals as a result of their own lifetime experience. The species-specific abilities form a basis, which is refined and developed in each individual. The adaptive capacity of an individual is not static; it continues to develop throughout an animal's life. At any one instant, it is dependent on the individual's internal state, as well as on changing environmental conditions.

In considering the welfare of individuals of more social species, where the environment also includes other members of its social group, it is necessary to re-evaluate the adaptive capacities of an individual as being related to the functioning of a social group as a whole. We suggest that welfare of a wider social group may be defined by adequate freedom to adapt to prevailing environmental circumstance *as a group* and that group welfare may be optimised, while the (objectively determined) welfare states of its individual members may vary over a considerable range, with nonetheless all members perceiving their own welfare state as being optimal.

The adaptive capacity of a group thus describes the set of (physical and mental) abilities with which a group of animals is naturally endowed. The species-specific abilities of each group member form a basis, which is refined and developed in each individual as a functional part of the whole. As with that of the individual, the adaptive capacity of a group is not static; it is dependent on the interactive functioning of group members as well as on changing environmental conditions.

Adaptive response

Welfare is in large part a function of an animal's ability to respond appropriately and in some adaptive way to its environmental circumstances. Adaptive responses are thus characterised by behavioural or physiological responses that enable an individual (or group) to react appropriately to both positive and potentially harmful (negative) stimuli (e.g. approaching a food resource or avoiding a potential danger).

Depending on internal (e.g. hormonal or developmental) and external changes (e.g. season) an individual may respond differently, even to the same stimulus at different times. While such different responses may all be adaptive, a distinct response may be more appropriate at a given juncture depending on prevailing internal and/or external circumstances. For example, foraging behaviour is clearly adaptive; still, during harsh weather conditions it might be more appropriate to seek shelter and to inhibit foraging behaviour. Thus, any meaningful assessment of the adaptive value of behaviour can never be done in 'absolute' terms but only in relation to prevailing circumstances.

Welfare

Welfare describes an internal state of an individual, as experienced by that individual. This state of welfare is the result of an interplay between the individual's own characteristics and the environmental conditions to which it is exposed and cannot be assumed to be the same for all individuals placed within a given environmental situation. Human determination of an animal's state of welfare is only as good as the observer's perception of the signals that the animal emits.

The welfare state of an individual represents a function of its adaptive functioning within prevailing environmental circumstances. For social animals, that environment includes other members of the social group or population; a separate assessment of welfare at the group or population level may thus be determined as the adaptive functioning of the group as a whole in response to a given welfare challenge. The adaptive functioning of a group is therefore the result of the characteristics of that group, as well as the environmental conditions to which the group is exposed.

At the individual level we assume that welfare is defined by the animal's ability and freedom to adapt to environmental conditions. But we recognise that individuals may show significant variation in their perception of a given status and their 'decision' about how to respond to that perceived status. Thus we may expect that even under identical environmental conditions, different individuals within a group or population may perceive or experience their welfare status differently.

Positive welfare

Positive (or good) welfare describes the state in which an individual, or group of individuals, has the adequate freedom to react to the demands of the prevailing environmental circumstances, resulting in a state that the animals themselves perceive as positive. With a growing emphasis on the importance of positive experiences (Fraser, 1993; Fraser and Duncan, 1998; Duncan, 2005), good animal welfare is not ensured by the mere absence of negative states (Knierim *et al.*, 2001; Duncan, 2005; Broom, 2010, see also Mellor, 2012) but requires the presence of positive affective states.

Negative welfare

In our view, and as a view increasingly expressed in the wider literature (e.g. Broom, 2006; Korte *et al.*, 2007; Ohl and van der Staay, 2012), negative or bad welfare status describes a state that the animal itself perceives as negative. Short-term negative welfare states, such as suffering from hunger and fear, serve as triggers for the animal to adapt its behaviour. They therefore serve a function. A brief state of negative welfare may fall within an animal's adaptive capacity, and would not necessarily require intervention. Welfare status is significantly compromised when an animal or a group of animals have insufficient opportunity (freedom) to respond appropriately to a potential welfare 'challenge' through adaptation by changes in its own behaviour (either where environmental challenges exceed the adaptive capacity of the animal or the opportunities available are inadequate to permit the animal to effectively express the appropriate adaptive responses).

Suffering

Suffering describes the negative emotional experience resulting from being exposed to a negative state of welfare. As noted earlier, short-term negative welfare states, such as hunger and even momentary fear, serve as triggers for the animal to adapt its behaviour. They may therefore fall within an animal's adaptive capacity and would not necessarily require intervention.

If an individual lacks the ability or the opportunity to react appropriately to any such negative stimulus, however, (for example, by escaping from a frightening situation), a challenge is created that may exceed the adaptive capacity of the individual. When this negative experience is profound (for instance, pain caused by an injury), we may consider the suffering acute however short-lived; when negative feelings are persistent over a protracted period, the situation may be considered to impose more prolonged suffering which may constitute a more significant welfare concern and may trigger some attempt at mitigation.

Ethical considerations

We will return to this issue of suffering in more detail below, but our brief mention here introduces an important additional dimension to our discussions to this point. Thus far we have been talking largely about what may constitute the welfare status of an individual or group of individuals. This welfare status may be bad, good or neutral, but is, *per se*, neither morally bad nor good. In our terminology, a welfare **problem** occurs if the adaptive capacities of an individual or group is being exceeded, but even so there is no implicit moral judgement.

However, as explored in more detail in Chapter 2, whatever our understanding of the underlying biology, any objectivity in analysis must ultimately be subject to interpretation and moral judgement when determining whether any given welfare status is or is not 'acceptable' to society. Whether or not the welfare status of an individual or group of individuals constitutes a welfare **issue** therefore indicates a value judgement by an observer or by society. The 'translation' of welfare assessments into management practice, and the way in which that management practice is viewed by society more widely, is markedly affected by public understanding and public attitudes (see Box feature).

In this chapter, therefore, we will also explore how biological understanding and an understanding of ethical values of society must be integrated into any 'universal' framework for welfare management.

The 'take-home message' from all of this is that assessment of what may or may not constitute a welfare issue is not simply dependent on objective biological evaluation of welfare status *per se*, but it must also take into account ethical dimensions and contemporary societal views. Societal and government policy on animal welfare and animal health should pay attention to the fundamental moral assumptions that underlie many animal-related problems (RDA, 2010). To help structure evaluation of the reported welfare

WELFARE PROBLEMS AND WELFARE ISSUES: LARGE HERBIVORES IN THE OOSTVAARDERSPLASSEN

Frauke Ohl and Rory Putman

The Ooostvaardersplassen is a nature reserve created on land reclaimed from the sea only 40 years ago in Zuidelijk Flevoland in the centre of the Netherlands. It is designated as an SAC (European Special Area of Conservation) and RAMSAR site. Management aims to create a mosaic of short-grazed and longer dry grasslands with areas of *Phragmites* reedbed fringing open shallow pools to provide suitable habitat for a diversity of wetland birds, such as greylag and barnacle geese, spoonbills, herons, great white egrets and others. Partly because it is such a wetland area (and as such it is not easy to use heavy machinery) and partly out of a desire to do things as naturally as possible, it was decided to try and impose appropriate vegetation management by introduction of free-ranging populations of large herbivores. Primitive breeds of both cattle and horses were introduced in 1983 and 1984 to try and mimic the effects of tarpan and aurochs, which might have occupied such habitats in past times. Because the area was becoming overgrown with elder, populations of red deer were also subsequently introduced in 1992.

The populations of large herbivores were left unmanaged (except for humane destruction of animals *in extremis*) and, as time went by, alongside a need to deliver the objectives for the site agreed under Natura 2000, emphasis shifted towards a secondary objective of minimum intervention. In the absence of natural predators or any human-imposed culling, populations of cattle, horses and deer increased rapidly; numbers soon exceeded the carrying capacity of the very limited area and there was significant mortality each year, especially over winter. The Oostvaardersplassen is bisected by the main railway line running to Amsterdam from the dormitory towns of Lelystad and Almere. Daily commuters soon noticed the number of dead and dying animals and there was a huge outcry. The relevant authorities therefore commissioned an independent review of management (ICMO, 2006; ICMO2, 2010).

Much of the focus of public concern was that animals were observed in very poor physical condition towards the end of winter – apparently starving and with very little flesh covering the bones. In reality, however, this is not necessarily a real problem: temperate ungulates have evolved to build up fat reserves over the summer and autumn and to draw on those same fat reserves over winter to make good any shortfall between energy intake and energy loss over that winter season (page 66). On its own, therefore, the observation of animals in poor physical condition towards the end of winter is not in itself a welfare problem (unless it is extremely acute). There were other more serious problems with management but, in reality, this particular concern was not itself a welfare problem. However, because the general public did not necessarily understand the dynamics of the system, it was certainly perceived as a welfare *issue*.

Image credit: E.M. Kintzen and I. van Stokkum

status of an individual or group of animals, a number of frameworks have been developed that may help to make explicit, structure and analyse moral issues in policy (see for example Beekman *et al.*, 2006; Mepham *et al.*, 2006; RDA, 2010). The application of such assessment models is not restricted to analysing practical questions of the morally 'right' action, but aims more at allowing for better structured and more explicit discussions on fundamental questions related to the moral good.

As one example, the assessment model proposed by the RDA (2010) consists of two parts (see Figure 6.1). The left column of the framework is focused on the applied value assessments, such as the question of whether the killing of a group of animals during an outbreak of animal disease is justified. The right column addresses broader and more fundamental questions related to a specific question, such as moral ideals on animal disease prevention. Ethical issues usually only tend to become explicit in a policy context if there is a clear need to deal with a specific problem, for example whether or not to shoot potentially suffering wild animals. The moral questions that underlie such a specific problem often remain unaddressed. The aim of the right column is to clarify these more fundamental questions in a way that is beneficial in

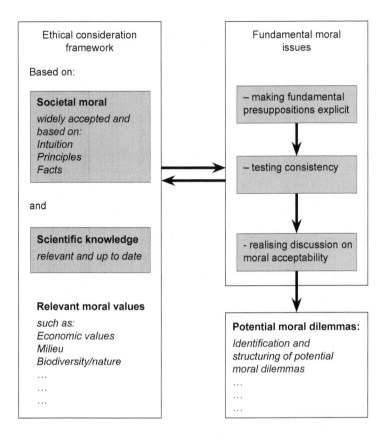

Figure 6.1 *Ethical framework.*

The left column of the framework is focused on value assessments, such as the question whether the killing of a group of animals during an outbreak of animal disease is justified. The right column addresses broader and more fundamental questions related to a specific question, for example whether or not to shoot potentially suffering wild animals. The aim of the right column is to make more explicit these more fundamental questions in a way that is beneficial in addressing the current dilemma, but also in drafting future policy.

addressing the current dilemma, but also in drafting future policy. The model or framework builds on the idea of ethics as a reflection process in which one strives for an equilibrium of a number of moral elements, including intuitive judgements, facts, principles and moral ideals (cf. Rawls, 1971; van der Burg and van Willigenburg, 1998).

The model starts with the intuitive judgements, i.e. the first impressions people may gain in a specific case. Intuitive answers often come to the fore in discussions about animals. In the search for a reflective equilibrium, these intuitions are considered to be a moral marker. They usually indicate that there is a moral problem or a question at stake. Still, such intuitions are only the start of an ethical reflection: a first reaction can be valuable, but people may also be mistaken due to lack of knowledge or the lack of acknowledging others' interests. Therefore morally relevant facts and ethical principles play a key role in the ethical consideration. The principles and relevant facts are employed in critical reflection upon intuitive judgements. This requires a critical reflection of the similarities and differences between first impressions, principles and facts. If it turns out that there are inconsistencies, one has to evaluate underlying causes and examine which parts are in need of modification. Once coherence is achieved between the intuitive judgements, the principles and the facts, a reflective equilibrium has been reached.

The assessment model introduced here is not a decision-making tool that functions as a one-size-fits-all approach for practical ethical dilemmas; instead, it should be clear that such an ethical framework is drafted in order to identify the relevant ethical questions and potential moral dilemmas rather than to yield straightforward management or political solutions.

Of course, as we have already highlighted, moral frames differ markedly between cultures and thus differ between societies, as well as showing a range of 'attitudes' even within one given society. Our assessment of what is acceptable must therefore be framed within a proper understanding of what the majority of members of our own contemporary society believe and accept, although this implies an additional responsibility to inform and educate wider society so that reactions are not simply based on untutored and unreflected intuitions (see e.g. Ohl and van der Staay, 2012; Meijboom and Ohl, 2012).

7

SHOULD RESPONSIBILITY FOR ANIMAL WELFARE VARY WITH CONTEXT?

There has been much debate in the more philosophical literature about human responsibilities to animals and the moral value of animal life. A relevant part of this literature recognises animals as having moral status: that is, as being an entity (a being) towards which we can have moral duties (Warren, 1997). However, there is a diversity of arguments that underlie this recognition of some moral standing of animals (see again Chapter 2 for more detailed review of this diversity and the practical implications).

This broad acknowledgement of animals as having moral status, however, appears not to result in one broadly shared view on how we should treat them. As an example, current legislative provisions for the management of animal welfare are, in general, closely tied to context, such that there may be a clear legal distinction between responsibilities defined towards farm animals, laboratory animals, companion animals, closely managed wildlife, and truly wild animals experiencing little management input (for an overview see, for example, Ausems, 2006; Vapnek and Chapman, 2010).[6]

Partly as a reflection of that legal distinction, some authors recognise a similar distinction in relation to context in terms of our moral responsibilities for animals and the way those responsibilities might be discharged (as, for example, responsibilities for kept versus non-kept animals; Swart, 2005;

---

6 As far as we are aware, the Netherlands and now Norway are the only countries in Europe which actually impose a legal obligation on *all* citizens 'to take responsibility and provide the necessary care for animals that need help' – whether wild or managed/kept (Dutch Animal Health and Welfare Act (GWWD), 1992 paragraph 36 subsection 3).The Norwegian Animal Welfare Act (January 2010) states: 'Animals have an intrinsic value which is irrespective of the usable value they may have for man. Animals shall be treated well and be protected from danger of unnecessary stress and strains. Anybody who discovers an animal which is obviously sick, injured, or helpless, shall as far as possible help the animal. If it is impossible to provide adequate help, and the animal is domestic or a large wild mammal, the owner, or the police shall be alerted immediately'.

Swart and Keulartz, 2011). However, we might question, after Webster (1994), whether our responsibility for the welfare of a mouse or a rat should really be any different if that animal is a pet, an experimental laboratory animal or a pest. We will argue that within those legal provisions that recognise an animal as having moral status (Ausems, 2006; Vapnek and Chapman, 2010) and which consider this moral status to result in certain responsibilities towards the animal, the *moral* duty to have regard for the welfare of any animal cannot be context-dependent, while the question of whether or not we should intervene in an given instance may depend on practical or economic constraints but not on any principal moral difference in obligation towards the individual as such (Ohl and Putman, 2014c).

If we wish to try further to justify such a claim for a universal (moral) duty of care, we may argue that at least within a Westernised context, all animals (whatever their status as kept or non-kept) are, to some degree, influenced by human activity – whether to a greater or lesser extent, and whether deliberate or incidental. Closely managed animals in whatever context (farm animals, laboratory animals, companion animals) have their whole environment controlled by human agency, but even for free-ranging or apparently wild animals, their habitat and movement patterns are affected by human land use and land management. Whether or not directly managed by man, wild animals suffer significant impacts from our activities in loss of habitat to agriculture, forestry or urban sprawl, loss of connectivity of habitats through the proliferation of road systems or other transport infrastructure, impacts on food quality, etc; animals are regularly killed and injured on our roads. Many wildlife populations are also directly managed by humans (for control, recreation or conservation) by culling or through use of other, non-lethal, control measures such as translocation, imposed contraception, and so on. Thus, even in terms of those philosophical considerations which base responsibilities for animals on their relationship with humans (e.g. Palmer, 2010; Swart and Keulartz, 2011), consideration of the effects of human activity on welfare might be expected even for apparently free-ranging wild animals.

In practice, however, such separate 'justification' is unnecessary if we simply argue that in sharing our planet with other living beings that have some inherent or intrinsic value we have a moral responsibility to consider their welfare. According to Regan (1983), animals, like humans, are subjects-of-a-life. Even though they are not moral agents, he claims that they have subjective experiences, can experience the quality of their lives, and enter into and maintain relationships with others. Thus, an inherent value is ascribed to an individual animal on the basis of it being a subject-of-a-life. Taylor (1986) argues that all organisms have 'inherent worth' because as 'teleological centres of life' they have a good of their own, while Rollin (2011) ascribes an intrinsic value to animals because 'what happens to an entity matters to it even if it does not matter to anyone or anything else. Because it is capable of valuing what happens to it, either in a positive or negative way, such valuing is inherent in it.' (For further review, see Chapter 2; also, for example, Cohen *et al.*, 2009, especially pages 345–349).

But if we do accept that, then the other side of that same argument is that the current compartmentalisation of legislation and responsibility makes no sense at all – and given a recognition of a strong unifying biological basis for what constitutes and contributes to animal welfare, we should be seeking methods of assessment of welfare status and requirements of welfare management which transcend these traditional compartmental boundaries so that they may be applied to any species and in any context – even when applied to truly wild, free-living animals (Ohl and Putman, 2014b).

Responsibility for action?

In effect, although the basis for the belief may differ (see Chapter 2; Appleby and Sandøe (2002) also Swart and Keulartz (2011) offer comprehensive reviews), most modern philosophers accept that all animals have some 'inherent value', or at least that humans have a wide-ranging moral duty towards them. What *does* appear to change with context is the degree of obligation (or requirement) to take action.

Swart and Keulartz (2011) link this responsibility to context in presuming that responsibility for addressing some compromise of welfare status is higher in animals that are more closely managed by man (or free-ranging animals more heavily impacted upon by human activities) because we are more responsible for providing all resources for more closely managed animals (farm or companion), which have less freedom and fewer opportunities to respond 'naturally'. They also suggest we should acknowledge a greater responsibility for wildlife species where humans are demonstrably responsible through their actions, either for the compromise to welfare status in the first place and/or for the restricted opportunities available to those wildlife animals to perform appropriate adaptive behaviour (perhaps because of human impacts in restricting habitat diversity).

We would argue that this negates the initial construct of an equal moral duty to all animals, which is universal and independent of the context of the human–animal relationship. We would, however, suggest that within that subset of cases where we may feel it appropriate to intervene, the obligation to take action is constrained by the actual practicalities of intervention and the availability of (realistic) mitigation measures to effect a change in welfare status. If there are no practical mitigation options possible then, almost by definition, this must lessen the obligation to take action.

However simple this last constraint may appear, it opens an additional debate about avoidable versus unavoidable suffering and, furthermore, whether our duty of care should be restricted to the prohibition of 'avoidable' suffering or should include the enhancement of positive welfare status.

Differing dichotomies in relation to animal suffering

In this context, suffering should be interpreted in a broad sense and, after Dawkins (1990, 2008), we here presume that suffering can result from experiencing a wide range of unpleasant emotional states such as fear, boredom, pain and hunger. While short-term exposure to such negative states/emotions is accepted as a necessary part of any mechanism triggering appropriate adaptive behaviour, if an animal continues to be exposed to such states over a prolonged period, then it is clear that the situation exceeds the individual's adaptive capacities and thus does constitute a welfare concern.

As a principal framework, we would suggest that the suffering of animals that is technically avoidable should be considered (morally) unacceptable. However, there may be constraints on intervention that are posed purely by practicalities of intervention or mitigation. Effective intervention in the lives of free-ranging or wild animals may simply not be feasible; for example in a farm context, if livestock animals are kept outdoors (which may itself be warranted in terms of promoting positive welfare in other respects), they may be exposed to extreme weather that may cause transient suffering. Sudden changes of weather are neither predictable nor controllable and occasional suffering (in the sense of short-term 'unpleasant emotional states') is *unavoidable* when animals are kept outdoors.

Other constraints may be posed by human convenience or other human interest, for example efficiency of animal production in an agricultural context. We suggest that *necessary* suffering in this context should be understood as that required to achieve some anthropocentric objective (e.g. animal experimentation, food production or pest control) or animal-centred goal (e.g. in veterinary practice). In such cases, there may be mitigation options that are not implemented for reasons of efficiency or economics. Here, any suffering resulting from non-intervention may be considered avoidable (in theory) but *necessary* (because of distinct subjective/individual human interests).

Cutting across these dimensions of necessary or unavoidable suffering, we must recognise that even science-based, operational definitions of animal welfare and suffering will necessarily be influenced by societal mores. Thus, in addition to determining whether considering an animal's suffering may be avoidable versus unavoidable, or necessary versus unnecessary, it is appropriate also to evaluate along a third axis, based upon what is considered by a wider society to be (morally) *acceptable* or *unacceptable* (RDA, 2012). Of course, moral frames differ markedly between cultures, and thus differ between societies, as well as showing a range of 'attitudes' even within one given society. Our assessment of what is acceptable must therefore be framed within a proper understanding of what the majority of members of our own contemporary society believe and accept, although this implies an additional responsibility to inform and educate that wider society so that reactions are not simply based on untutored and unreflected intuitions (see e.g. Ohl and van der Staay, 2012; Meijboom and Ohl, 2012).

To intervene or not to intervene?

Following this line of reasoning, the obligation to take action in response to a perception of suffering, or in an attempt to promote more positive welfare of an individual or population, is constrained by what may be considered necessary or acceptable suffering, or by the actual practicalities of intervention and the availability of (realistic) mitigation measures to effect a change in welfare status.

If we accept the more dynamic concept of what constitutes welfare presented in Chapter 3 (and rehearsed, for example, by Duncan and Fraser, 1997; Broom, 2006; Bracke and Hopster, 2006; Korte *et al.*, 2009; Koolhaas *et al.*, 2010; Ohl and van der Staay, 2012): that a welfare problem or concern arises only when an animal or a group of animals have insufficient opportunity (freedom) to respond appropriately to a potential welfare 'challenge' through adaptation by changes in its own physiology or behaviour, that then implies that biologically we have an obligation to take action to address potential welfare problems only in those situations where the animal cannot adapt appropriately by changes of its own physiology or behaviour (sufficient to bring about appropriate environmental adjustments required to restore its positive welfare status).

Yet from our deliberations here, we must recognise that even within such a 'subset' of cases, if there are no practical mitigation options possible, then almost by definition, this must lessen the obligation to take action. Thus the obligation to intervene (in any context, wild or closely managed) depends primarily on whether or not there are practical (and economically feasible) options available for intervention (regardless of whether such intervention is in the interests of avoidance of suffering or positive, proactive enhancement of general welfare status).

It is clear that even in such cases, some interventions are mistaken or misguided and run the risk of resulting in greater harm, whether to the animal directly concerned or, by their consequences, in other parts of the biological system to which it belongs (in each case a casualty of unforeseen and unintended consequences). As a refinement of the responsibility for intervention therefore, we may suggest, that our obligation to intervene (where it is practical to do so) should further be restricted to those cases where we can be certain that our intervention will not result in greater harm (Sozmen, 2012; see also Donaldson and Kymlicka, 2011). [This perhaps avoids the obvious logical inconsistency of interfering within, for example, a natural predator–prey interaction (see, for example, Norton, 2005; Gamborg *et al.*, 2012).]

THE RISK OF UNINTENDED CONSEQUENCES

Rory Putman

I earn my living primarily as a wildlife biologist, with a particular interest in large ungulates and their management. Wherever I settle, word soon spreads of my 'expertise' and so, increasingly, local people have brought me the wildlife casualties they have encountered; even the police and the local vets would ask me to take on rearing or rehabilitation of wildlife brought to them after vehicle accidents on the road, or 'orphans' found and 'rescued' by well-meaning dog-walkers in the local woods. Too often the 'orphans' picked up by well-meaning members of the public and rushed to the local veterinary practice or welfare charity haven't been abandoned at all; simply when they are very young, they do not have the strength as yet to run after their mother, so that when she goes out to forage, she hides them in cover in a safe place, returning at regular intervals to suckle them. Nine times out of ten when the dog-walker finds an 'abandoned' fawn, the mother is not far away, probably even watching from the bushes. But each year, well-meaning folk pick up dozens of these 'orphans' and bring them to animal orphanages to care for. And by the time they are brought in, it is often too late to return them to where they were found.

Over the years, therefore, my wife and I ended up rearing countless of these supposedly abandoned juveniles of a variety of species, but especially deer – and among the deer, primarily roe deer. Over the years we have refined our approach to have greater and greater success in rearing the little fawns, while at the same time avoiding letting them become over-tame – because if they become too tame then it is difficult to release them successfully back to nature; further, having lost their fear of humans, males, in particular, can become a nuisance and once they have fully developed their antlers, potentially even dangerous. (Each year a number of human deaths or at least serious injuries result from aggression from habituated male roe deer).

I know of relatively few studies of survival after release. Studies of translocated adult white-tailed deer, black-tailed deer or mule deer in the US suggest a very low success rate – less than 10 percent – but these are generally studies of adults being live-captured and translocated into new and unfamiliar areas, not fawns. All roe reared by me were ear-tagged and monitored for a period after release and we enjoyed far greater rates of success: in excess of 80% of our hand-reared fawns survived for at least 2 years after release. One might argue that this was a huge success story and indeed a welfare success. Faced with a beautiful tiny kid in a cardboard box, it is hard to harden your heart, but a part of me began to question the implications of our apparent success. *Our* animals survived of course, but at what cost?

Roe deer are pretty ubiquitous throughout Europe and one of the most abundant of the deer species. While we had a reasonably good success rate in rearing these rescued fawns, we were inevitably releasing them when subadult into an environment pretty well-saturated with other roe deer. Although not strictly territorial, roe deer are not especially tolerant of competitors in their home range. If the environment was already filled to capacity, was perhaps another roe deer being pushed out to die for every healthy kid we successfully reared and released? The wild deer were of course unmarked so it was impossible actually to resolve such a question, but it illustrates nicely the potential risk of intervention within a balanced system and the risk of causing greater harm as a result of that intervention. For me now, when faced with a fawn in a cardboard box, the difficult issue ethically is that I do not know what was the effect of such releases on others in the wild roe population already present where my animals were released.

Image credit: Rory Putman

We may summarise much of this 'thought-process' in debate over whether or not a given situation demands some intervention (and thus our human responsibility for action in a given situation), in the decision-making framework presented in Figure 7.1.

Finally, as well as discussing whether or not one should intervene at all in any given situation of potential animal suffering, we may also discuss what might be the appropriate form of intervention in cases where we do feel it is our responsibility to do so. While we disagree with Swart (2005) and Swart and Keulartz (2011)'s contention that duty of care is itself dependent on

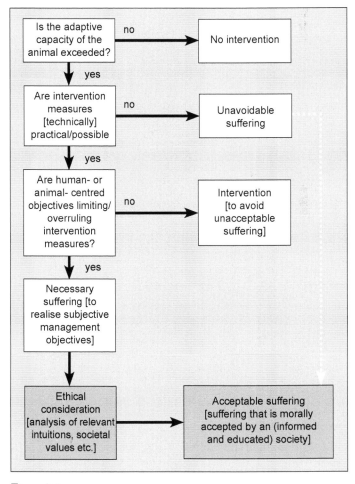

Figure 7.1

context, Swart (2005) argues convincingly that the *form* of that intervention (if intervention is justified at all) might indeed be more dependent on context – such that for wild animals, welfare is more likely to be affected by 'non-specific care' focused on maintaining and developing the natural environment of the wild animal so that it can lead a natural life. As the counterpart of non-specific care, Swart suggests 'specific care' for animals kept by humans, including domesticated animals. These animals are entrusted to our care or are placed in human environments, and are therefore much more dependent on us for their welfare and needs. Although the two different types of care exclude each other to a certain extent, they do not rule each other out completely, since there may be good reason to provide specific care when an individual wild animal happens to find itself in acute distress (Swart, 2005; Swart and Keulartz, 2011).

While we would decouple this construct from an initial contention that the responsibility for care is in the first place dependent on context, this suggestion that the form of intervention must be more closely related to circumstances seems entirely appropriate. Swart's deliberations also 'fit' nicely with the concept of what managers of wildlife populations might concern themselves with, in relation to proactive measures aimed at improving the welfare status of wild animals through general environmental enhancement, rather than a simple focus on reducing suffering caused by culling or other population control measures (e.g. Cockram *et al.*, 2011).

In conclusion, we would suggest that:

◊ the moral duty of care is absolute and independent of context.

◊ the requirement to intervene should be based on biological assessment of whether or not sufficient opportunities exist for the animal or animals to respond adaquately to a potential compromise of welfare status through appropriate and adaptive changes in its own behaviour (sufficient to bring about appropriate environmental adjustments required not only to avoid suffering, but to restore its positive welfare status). The requirement for intervention in such cases is also constrained by the physical possibility/impossibility of any effective mitigation (avoidable versus unavoidable suffering).

◊ intervention is further constrained by considerations of human interest in the animals concerned (thus necessary versus unnecessary suffering).

◈   there is in addition a moral dimension in co-determining the scale of required intervention in relation to societal norms of what may be acceptable versus unacceptable suffering.

We believe that a construct cuts across the rather artificial distinctions we have previously and traditionally used between kept and non-kept animals (it obviates the need for such artificial and arbitrary distinctions) and replaces these with a more robust and functional construct applicable in more general terms. In effect, the whole neatly defines a recognition of welfare concerns (in all contexts) as lack of opportunity to respond appropriately to environmental challenges, and offers a simple concept of appropriate solutions in terms of providing the opportunities which enable the animal(s) to react appropriately to such challenge. It provides a recognition that we (humans) have an obligation to take action only if an animal is (for whatever reason) unable to respond appropriately and effectively through its own adaptive behaviour; a recognition that obligations are further constrained by actual practicalities of what is possible in terms of mitigation.

APPLYING THE ASSESSMENT FRAMEWORKS TO A CASE STUDY: SUPPLEMENTARY WINTER FEEDING OF DEER

Frauke Ohl

The practice of providing supplementary feeding for wild deer over winter is a controversial one and provides a good illustration of the need to resolve exactly objectives and welfare implications of such supplementation because assessment of the 'acceptability' of such intervention may depend very much on objectives (and see Figure 7.2). Briefly, such food supplementation over winter may be provided in an attempt to offset a perceived shortage of natural food during the winter, thus assisting animals to maintain body condition and/or to reduce overwinter mortality.

Alternatively, food may be provided for completely unrelated objectives in a belief that it will improve antler mass and trophy quality of mature stags, as a device to try and heft stags more closely to a given property (and stop them wandering further afield), or as a diversionary tactic to draw wintering animals away from areas which might be sensitive to sustained heavy impacts (unfenced forestry or vegetation of high conservation value, where excessive impacts might be damaging).

We can use the assessment frameworks of Figures 6.1 and 7.1 to explore the possible justification for intervention and the form of any such intervention as in the figures below.

Since the primary objective has significant implication for the aspects we need to evaluate our considerations, as well as the route of exploration, it is essential to define that objective (see Figure 7.2). If winter feeding is being considered, or is actually being undertaken to ameliorate a perceived welfare problem/welfare issue, we first have to ask whether there *is* a welfare issue in the first place (see Figure 7.3).

Here, this resolves to:

- Is there a shortage of winter food or is the seasonal decrease in food availability within the animals' adaptive capacity (i.e. within limits for which it/they possess appropriate physiological or behavioural adaptation)?

- Does seasonal decrease in food availability exceed the adaptive capacity of some individuals or that of the group?

If a welfare issue is perceived, related biological questions might also be:

- Is it food that is limited at all (or shelter)?

If the answers to both of these suggest that supplementary feeding may be appropriate and an effective way of addressing a real welfare problem (that the welfare of the deer *is* genuinely compromised by lack of natural food over winter and provision of appropriate supplementation might improve welfare overall), then we should also consider the questions:

- Is supplementary feeding ethically acceptable?

  [*In some countries such as Austria and Germany it is actually compulsory, but to many people it is not ethically acceptable because it is felt that it is in some way 'interfering with nature' – and in the Netherlands, for example, as well as in many parts of the United States, it is expressly forbidden*].

- *Does it actually achieve what it is supposed to achieve in terms of maintaining welfare or reducing overwinter mortality?*

- *Does winter feeding affect animal welfare at the individual level or at the level of the group?*

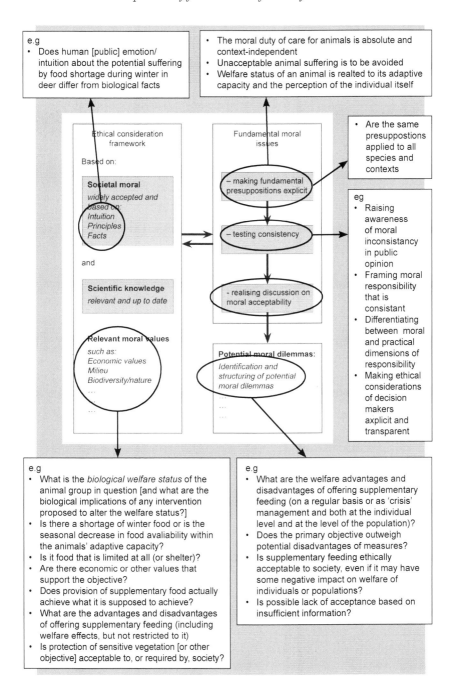

Figure 7.2 *Assessment of the biological and ethical dimensions associated with winter feeding of deer.*

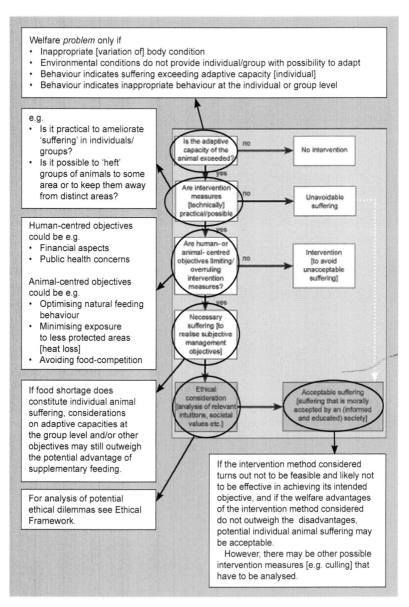

Figure 7.3 *Assessment of the necessity for providing supplementary winter feeding for wild deer from a welfare perspective: when is intervention justified? (adapted from Ohl and Putman, 2013c)*

8

ASSESSMENT OF WELFARE STATUS AT INDIVIDUAL OR GROUP LEVEL

Given everything discussed so far, how may we set about actually assessing the welfare status of animals, whether individually or collectively? In practice, there is a huge range of different protocols which have been developed over recent years in order to try and determine some measure of welfare status. Inevitably such methods have, in the main, been developed for application to farm animals, pets or (separately) laboratory animals, and many are not immediately transferable to animals which are less closely managed. This is because, as a short-cut, in systems where environmental conditions are largely controlled by human agency, many of these methods focus less on the actual observed welfare of the animals themselves but are instead more concerned with physical features of the environment in which the animals are maintained (e.g. in farm livestock the length of stalls, feeding and drinking facilities, space allowance, access to pasture, etc.), in relation to whether these meet certain minimal criteria *presumed* to be sufficient to confer adequate welfare status.

Further, as we have noted, many such protocols have been based on the avoidance of negative states ('avoidance of suffering'), with approaches to the assessment of welfare status commonly based on the supposition that positive welfare merely implies freedom from such negative states. Once again, the Brambellian tradition has established a certain conservatism amongst practitioners and frequently assessments of the 'welfare' of farm or laboratory animals barely consider the animals themselves but focus instead on the physical features of their controlled environment and whether or not those meet the specifications currently stipulated by legislation and considered to deliver the Five Freedoms. Except within the context of research directed towards determining such protocols and informing legal requirements, where assessments are made of the welfare status of individual animals, these are usually undertaken by a veterinarian during routine inspections, and are largely based on assessment of health and physical condition.

However, as we have already noted, more recent commentators on welfare suggest that society's responsibility may not be limited simply to the avoidance of negative states in animals that are directly managed by humans, but that there may also be an implicit responsibility for active facilitation of positive welfare-states (see, for example, Dawkins, 2008; Yeates and Main, 2008; Ohl and van der Staay, 2012). In addition, it is generally accepted that welfare should not simply be assessed as a single static measure at one given instant of time, but as a more dynamic, interactive state reflecting an animal's capacity to adapt to environmental challenge by changes in its own physiology or behaviour. Crucially, therefore, our assessment of welfare should not simply consider the status of any individual at a given moment in time, but needs to be integrated over the longer time periods required to execute such change.

The necessity of such a dynamic concept for the assessment of animal welfare becomes even more apparent when applied to free-ranging, wild animals. For example, the natural dynamics of natural environmental processes in seasonal environments result in seasonal changes in food availability for populations of grazing animals such as wild deer, resulting by definition in periods of hunger. In relation to the considerations presented in the Box feature in Chapter 6, for example, we should recognise that temperate grazers have evolved to be able to cope with changing conditions by building up fat reserves in periods of food abundance that supplement the shortfall in periods of more restricted food intake (e.g. red deer, *Cervus elaphus* (Mitchell *et al.*, 1976; Kay, 1979; Kay and Staines, 1981); white-tailed deer, *Odocoileus virginianus* (McEwen *et al.*, 1957; Short *et al.*, 1969; Moen, 1976, 1978); black-tailed deer, *O. hemionus columbianus* (Wood *et al.*, 1962), reindeer, *Rangifer tarandus* (McEwan and Whitehead, 1970; Leader-Williams and Ricketts, 1982); free-ranging ponies (Pollock, 1980; Berger, 1986; Mayes and Duncan, 1986; Gill, 1988, 1991; Burton, 1992)). Thus, although society may consider protracted periods of presumed hunger to be a welfare issue, in practice it does not constitute a direct problem for the animal itself, and welfare will thus be compromised only if food restriction persists beyond their adaptive capacity (i.e. extreme depletion of their body fat reserve).

Perhaps most important for our deliberations in this section is that it has now been generally (if not universally) accepted that an individual's state of welfare is not a fixed parameter that can easily be assessed by external indicators or 'scored' by some external and fixed criteria, but is coloured to a large extent by the animal's own perception of its status. And within any given population, individuals may well vary in their internal perception of and response to the same 'absolute' status (Chapter 4).

All such considerations challenge the more functional protocols developed for assessing welfare status, even for closely managed animals, especially when these are based entirely or almost entirely on assessment of environmental parameters. It becomes clear that if we accept that welfare is defined by an ability to adapt and respond to environmental challenge in an appropriate way – (and that both positive and negative welfare states are a function of the actual adaptive capacities of the individual animal and the opportunity it has to express those responses) – then assessments based exclusively on the simple measurement of environmental parameters will prove inadequate, and the assessment of welfare must instead be based primarily on detailed observation of the physiological condition and behavioural responses shown by individual animals over time. Animal-based measures are, in this sense, more direct measures of welfare than simple assessment of those environmental parameters considered *a priori* to be determinants of good welfare because they attempt to assay the state of the animal itself.

However, the recording of some of the animal-based parameters is difficult and demands considerable resources – and even when they are recorded, the results may be difficult to interpret (Johnsen *et al.*, 2001). Many physical parameters of health (body condition, disease status) are relatively easily measured and, most importantly, measured values can be assessed against (usually) well-established normal ranges (Duncan and Petherick, 1991; Mench, 1993; Sejian *et al.*, 2011). As we have noted, however, a full assessment of animal welfare involves the consideration of a wide range of factors in addition to those purely reflecting physical health (Duncan, 1993; Moberg, 1993), and for many of these factors it may be impossible or inappropriate (given the extent of individual variation) to establish normal ranges (see again Sejian *et al.*, 2011).

In practice, whatever methodologies are adopted will be determined in part by practicality and constraints of time facing those undertaking the assessments, as well as the primary purpose of the assessment. As noted by Johnsen *et al.* (2001), much will depend on the reasons underlying a particular assessment and what the measurements are intended to deliver. If the goal of a given assessment is to evaluate a given production system across farms, or to certify that the conditions of a housing system are as they are claimed to be, then it may be sufficient simply to examine environmental parameters. However, if the goal is to reveal welfare problems at herd level and to provide advice of how to improve welfare on the farm, then records of environmental parameters must be combined with records of animal-based welfare parameters (Whaytt et al., 2003).

Assessments based on environmental conditions

Examples of welfare assessments primarily focusing on housing systems and management are the Animal Needs Index (ANI) and the Freedom Food Scheme of the (English) Royal Society for the Prevention of Cruelty to Animals (RSPCA) – both directed primarily at livestock production systems. The ANI system is based on assessment of a number of key components of husbandry (freedom of movement, social environment, condition of flooring, indoor climate and stockman's care), while the Freedom Food Scheme is based on delivery of the Farm Animal Welfare Council's (FAWC) Five Freedoms (see Chapters 1 and 3; FAWC, 1994) and involves a systematic survey of environmental conditions on the farm, but does not include indicators actually related to the animals themselves, or the quality of stockmanship (Main et al., 2001).

Animal-based assessment of welfare status

In general, we can break down the various animal-based approaches used in assessment of animal welfare into three main categories (although these categories are non-exclusive): (1) naturalistic, (2) functional and (3) those based on subjective experience (Duncan and Fraser, 1997). Naturalistic approaches are based on the premise that the welfare of any animal depends on it having the freedom to perform its natural behaviours and live as natural a life as possible. This approach has the advantage that it has an intuitive appeal and corresponds neatly with public understanding of what might constitute good welfare; however, it actually idealises 'naturalness' and does not recognise that many animals have a good capacity to adapt to more artificial environments (Duncan and Fraser, 1997; Sejian et al., 2011).

More functional approaches to welfare concentrate exclusively on the biological functioning of the animal and consider that welfare status is related simply to good biological functioning and the satisfaction of primary needs. The biggest advantage of such assumption is that the variables involved are substantive and fairly easily measured (Sejian et al., 2011); in many ways, however, it is too 'functional', especially if we believe that welfare is, to some significant part, determined by how an animal feels about its own status. On the other hand, such feelings or emotions are hard to define and assess, and this probably accounts for the resistance of the biological-functioning school of thought to the idea of including within an assessment of welfare any element of the animal's own self-assessment, restricting consideration instead to things that can be clearly and objectively measured. 'Science should be objective... and measuring biological functioning [when assessing animal welfare] ensures that objectivity' (Bracke, 2007).

Many continue to argue that the psychological well-being of the animal is fundamental to its welfare and that some recognition of this helps to overcome the problems implicit in any assessment of the variation in individual perception of, and response to, a given welfare status. (Some methodologies embrace a mixture of 'emotional' cues with those related to biological function). If, in this way, it is feelings that govern welfare then, in spite of the obvious difficulties of measuring feelings, it is feelings that should be assessed (Duncan, 1996): that said, the feelings and emotions of animals are extremely hard to assess in any consistent and objective way.

In the pages that follow we will attempt to review some of the more formal and informal measures available for animal-based assessments of welfare status, whether based on biological functioning or some assessment of subjective experience. Some of these measures overlap (in the sense that measures which accurately and objectively record stress are also some reflection of the animal's own perceived emotional experience).

Physiological approaches to measurement of welfare

In general, physiological approaches to the measure of welfare have focused on the concept of measuring levels of stress experienced by individuals based on the belief that if stress increases, welfare decreases (Dantzer and Mormede, 1983; Dantzer *et al.*, 1983; Moberg, 1985). However, there are a number of problems with such an approach. Short-term stress responses are an inevitable part of the process triggering an adaptive response from the animal and thus may be functional in maintaining a longer term positive welfare status. In such analysis, a more relevant measure might be evidence of chronic and 'traumatic' stress, something which is not trivial to differentiate by means of physiological measurements from acute stress (McEwen *et al.*, 1992; deKloet *et al.*, 2008a, 2008b).

Physiologically, stress is characterised by an activation of the Hypothalamus–Pituitary–Adrenocortical (HPA)-axis, resulting in the release of cortisol into the blood (Selye, 1950). The majority of approaches to the measurement of stress therefore consider stress levels reflected in the identification of elevated cortisol levels (although a variety of other blood or tissue parameters have also been considered, see below). However, chronic stress can result in a blunted HPA response and, as a consequence, a *reduced* release of cortisol in response to acute stress (McEwen *et al.*, 1992). In other words, based on low levels of peripheral cortisol alone, it is impossible to discriminate between absence of stress or a chronically stressed status.

Where such methods are applied to the assessment of stress within living individuals, it is clear that collection of blood or tissue samples from free-ranging wild animals is technically complex and, here especially, elevated levels of stress-related products may simply be due to the acute stress associated with capture and sampling. Furthermore, for many of the other parameters assessed, such as lactic dehydrogenase (LDH-5), muscle glycogen, bilirubin, etc. (e.g. Jones and Price, 1990, 1992; Price and Jones, 1992; Bateson and Bradshaw, 1997; Cockram *et al.*, 2011), there is some difficulty in separating the effects of chronic or acute stress from those which may simply be associated with vigorous or prolonged physical exertion. In addition, interpretation of results may also be made more complicated due to circadian and ultradian rhythms of hormones and other factors such as activity levels and the reproductive state of an animal.

When levels of cortisol are examined in sources that reflect accumulation of cortisol excretion over time, such as hair or dung, some of the confusion between measures of acute and chronic stress may be avoided (Sheriff *et al.*, 2011). However, here again we run into problems of interpretation in that chronic stress can result in a blunted HPA-response and, as a consequence, a *reduced* release of cortisol in response to acute stress and it is therefore impossible to discriminate between absence of stress or a chronically stressed status.

There may, however, be other non-invasive methodologies that may permit assessment of levels of chronic stress. When the HPA-axis is activated due to stress, this in itself causes changes in blood flow which may result in alteration of core and peripheral body temperature. A number of studies have shown that changes in the temperature of the periphery caused by acute stress consist of an acute and a chronic phase. During the acute phase, the blood is redistributed in the animal's body, resulting in elevation of core temperature and a decrease in temperature in the periphery. This rapid decrease in temperature of peripheral regions following stressful events has been described in several studies (Vianna and Carrive, 2005, Stewart *et al.*, 2008, Edgar *et al.*, 2013). For example, Stewart *et al.* (2008) have reported a significant drop in eye temperature following disbudding without anaesthetic. This initial response to acute stress probably serves to protect the animal from losing blood in case of an injury and to increase perfusion pressure. The second phase that follows includes a prolonged period of elevated temperature of the animal's surface which can last up to several hours. Such changes of core and peripheral body temperature can lead to the potential of levels of stress being determined in a non-invasive manner through appropriate infrared imaging (infrared thermography, IRT; Stewart *et al.*, 2005, McCafferty, 2007).

Although, the best site for measurements could possibly vary from species to species,[7] various studies have shown that the most suitable site to measure stress-induced changes in body temperature of animals is the eye because this is an area which is highly sensitive to alterations of blood flow (Stewart *et al.*, 2007; Valera *et al.*, 2012; Bartolomé *et al.*, 2013). According to a study by Dunbar *et al.* (2009), eye temperature of mule deer did not differ significantly from core temperature. It was also shown from a study conducted on ponies that their eye temperature measured by IRT was significantly associated with core body temperature measured by rectal thermometer and implanted thermal microchip (Johnson *et al.*, 2011).

In a study conducted on horses it was shown that IRT measurements were correlated to saliva and plasma cortisol concentrations after exogenous stress induction, suggesting a relationship with HPA-axis activation (Stewart *et al.*, 2005). A study by Bartolomé *et al.* (2013) in which IRT measurements were taken in horses before, during and after a competition suggested a similar physiological basis of eye temperature and heart rate. However, it was shown that peripheral body temperatures were also significantly affected by other factors, such as the age and the genetic line of the horses. The fact that the eye temperature remained elevated for at least three hours after the competition suggests that this increase could be long-lasting, even after the response to an acute stressor.

Unfortunately, whether and how core body temperature and surface temperature may change in response to *chronic* stress remains unknown. However, it is known from previous studies on laboratory rodents and humans that prolonged stress can lead to a significant increase of an individual's core temperature (Endo and Shiraki, 2000; Oka *et al.*, 2001). For example, the results of a study conducted on rats showed that repeated psychological exposure to stress led to an increase in core temperature for a prolonged period of time after the termination of the stressful situation (Endo and Shiraki, 2000). Similar findings have been presented in the study by Hayashida *et al.*, (2010) where repeated stress resulting from defeat in

---

7 For example, Edgar *et al.*, 2013 suggested that the best site to measure handling stress in domestic chickens was their comb, while a pilot study conducted on rabbits suggested that both the animals' eye and ear were appropriate sites for measurements (Ludwig *et al.*, 2007).

aggressive social interactions was found to induce chronic hyperthermia in rats.

Some pioneering work currently in progress at the University of Glasgow suggests that measurement of temperature at the surface of the eyeball may indeed have potential as an indicator of levels of chronic stress experienced (see Box feature).

CHANGES IN EYE-REGION TEMPERATURE MAY OFFER AN INDICATOR FOR CHRONIC STRESS

Rory Putman and Dominic McCafferty (University of Glasgow)

The body temperature of most birds and mammals varies according to endogenous circadian cycles and in response to changing environmental conditions. However, body temperature also responds to stressors that influence homeostasis. Notable changes occur in response to acute stressors that activate the autonomic nervous system. This can be detected as stress-induced hyperthermia (warming) in the body core and cutaneous hypothermia (cooling), as blood is diverted from the periphery to the main muscles and vital organs. These changes occur over a few minutes and may not persist long after stressor has ended. What is not yet fully understood is how body temperature may change over longer time periods (days/months) in response to prolonged chronic stress.

Preliminary studies on captive birds at the Institute of Biodiversity, Animal Health and Comparative Medicine, University of Glasgow, indicated that changes in peripheral temperature may vary with environmental enrichment, offering an opportunity of using surface body temperature as a measure of chronic stress state. Surface temperature is a useful measurement because it can be recorded non-invasively and remotely using infrared thermography (thermal imaging). The fur and feathers of mammals and birds insulate the skin and therefore bare skin areas or eye region temperature allow the best measurement of changes in peripheral temperature over time.

This technique has been applied in field studies of fallow deer (*Dama dama*) maintained in deer parks in different areas of the United Kingdom. In this study, eye-region temperature showed clear differences between populations once the effects of measurement distance, ambient temperature and sex had been accounted for (McKeon, 2016). Interestingly, while there were differences in median temperature, the greatest differences between populations were observed in the variance in eye-region temperature between individuals. This is not unexpected. If every individual in a given population is acutely stressed, then all animals within a population are going to respond strongly to that challenge, whereas if the stress imposed is comparatively mild, then different individuals may react more or less strongly and variation will be higher.

Variance in residual eyeball temperature in female deer in winter appeared to be strongly influenced by climatic severity (**environmental stress** – specifically, average temperature in January and February and the number of days of snow typically experienced in their particular geographic locality). Variance in eye-region temperature for both males and females was also strongly correlated with various aspects of population density both in summer and winter. Relationships were strongest with overall density of fallow deer and, in some instances, specifically with density of male or female fallow deer. Such observation implies that deer were showing some response to an element of **social stress**, possibly due to increased rates of social interaction at higher density, or related to competition for available food.

Image credit: Jackie Pringle

Behavioural approaches to measurement of welfare

Such techniques may in due course offer an alternative tool for assessing stress as an element of welfare but, in reality, many of these methods are technically complex and costly in application and perhaps more suited to research than to routine management. In addition, interpretation in relation to actual welfare status is always difficult, given earlier comments about the extent of individual variation in perception of and response to apparently identical stressors. It would appear that, for the most part, behavioural observations offer a simpler and more robust approach to assessing welfare from the animal perspective (Le Neindre *et al.*, 2004; Moura *et al.*, 2006). The behavioural choices animals make and the amount of stress shown when making those behavioural decisions may be quite revealing in terms of their status and perception of that status (Costa, 2003; Dawkins, 2003).

Thus in theory (and certainly in application to more closely-managed animals such as domestic livestock, laboratory animals, companion animals – even wild species in captivity in zoos or other menageries), behaviour may be scrutinised for the occurrence of persistent appetitive behaviour, indicating that an individual is seeking to express some adaptive behaviour in response to a given environmental challenge, but is unable to do so effectively – whether because it either lacks the correct adaptive response or because the constraints of its environment do not offer it the correct opportunity to express that behaviour effectively.

In addition, close observation may reveal persistent behavioural indicators of frustration (defined classically as the lack of some expected stimulus) or thwarting (again defined strictly as the animal being physically prevented from expressing some behaviour for which all internal and external stimuli are present). Responses to frustration or thwarting are characteristically expressed in rapid alternation of behaviour, performance of incomplete behaviours or redirection of behaviour to 'inappropriate' objects, as well as classic 'displacement activities' (for review, see Eilam *et al.*, 2006). When such 'deprivation' is maintained over long periods, animals may develop clear 'stereotypic' patterns indicative of persistent frustration (Insel, 1988; Eilam *et al.*, 2006).

Obvious examples of such behaviours are crib-biting in horses (Hothersall and Casey, 2012), tail-biting in confined pigs (Taylor *et al.*, 2011), feather-pecking in birds (Eilam *et al.*, 2006), and a variety of other self-mutilating behaviours in different species (Dantzer and Mormede, 1983; Fraser and Broom, 1990). In captive wild animals, repetitive motor rituals are common (for review, see Eilam *et al.*, 2006) which may be either **stationary** (movements executed without progressing in the environment) or **pacing** (progressing along fixed paths). Following Eilam *et al.* (2006), stereotyped pacing may be more prevalent in animals that range over large territories in the wild, while stationary movements are more typical of farm animals and pets.

In animals that are less restricted, environmental challenges that exceed the adaptive capacities of an individual are likely to result in both a lack of the behaviour necessary to fulfil actual demands (such as foraging: see Fraser *et al.*, 1997; Mellor, 2012) but also in a lack of expression of positive emotional states (such as comfort-behaviours). Positive affects or emotional states may include pleasure, comfort, contentment, curiosity and playfulness (Mellor, 2012), which suggests that regular observation of such 'positive' behaviour-types might argue for the individual(s) concerned being in a status that it perceives as positive. We should note, however, that the *absence* of positive indicators is not sufficient on its own to prove a negative welfare state.

Routine protocols of behavioural observations should therefore include positive indicators, such as play behaviour, and notice that the regular absence of play behaviour in young/adolescent individuals is likely to indicate high environmental pressure (Held and Spinka, 2011). In adult animals, play behaviour may be observed less regularly while, for example, active exploration and social- and self-grooming behaviour can be expected to be present on a regular basis (Crofoot *et al.*, 2011; Kikusi *et al.*, 2006). Further behavioural extremes may also be observed, such as high levels of intra-group aggression (Koolhaas *et al.*, 2010) or changes in group structures. All such changes not only indicate that prevailing environmental conditions are exceeding the animals' adaptive capacities, but are exerting chronic stress and will likely result in a further decrease in physiological condition.

Qualitative Behavioural Assessment (QBA)

Formal and objective (scientific) evaluation of an animal's welfare status through behavioural observations of this sort is time-consuming and requires somewhat specialised training. In an attempt to short-circuit such assessment, Wemelsfelder and co-workers developed the Qualitative Behavioural Assessment (QBA) approach, which is based on distilling and combining the subjective judgements of supposedly independent but untrained observers, each of whom is asked to offer, on a 5- or 10-point scale, their assessment of an animal's 'well-being' based on subjective appraisal of their general behaviour (Wemelsfelder and Lawrence, 2001; Wemelsfelder, 2003, 2007). This approach attempts to integrate the different aspects of an animal's interaction with its environment – not by offering detailed description of what the animal does, but rather as an assessment of how does it, using descriptors such as 'calm', 'anxious', 'timid' or 'confident' (Wemelsfelder and Lawrence, 2001).

To some extent, this offers a type of behavioural profiling akin to our earlier recognition that different individual animals may have different 'personalities' in the way they respond to different situations, or different 'coping styles' (page 32). However, Wemelsfelder and co-workers extend this approach in suggesting that such profiling may also assess the animal's current subjective experience and thus its actual welfare status (Wemelsfelder, 1997). They point out that animal professionals (producers, veterinarians, animal handlers) routinely use subjective descriptors to discuss and manage their animals' health and welfare state and that potentially such assessments therefore provide a suitable practical welfare assessment tool, as long as such assessments were found to correlate well with conventional measures of welfare.

Using a system of Free-Choice Profiling (FCP), whereby observers do not work with pre-fixed lists of descriptors, as are commonly used in animal temperament and personality studies, but generate their own descriptors based on close observation of animals in various test situations, Wemelsfelder and colleagues showed groups of untrained observers (although familiar with farm animals) video clips of animals in various settings and asked them at the end of each clip to write down adjectives which they thought adequately described how the animals had behaved (Wemelsfelder, 2007). Having thus generated a list of descriptors for the observed animals' expressions, observers would then be asked to watch the same video material again, and use their personal terminologies to quantitatively score the intensity of perceived expressions, for example how shy or lively they thought an animal was. Observers were asked to stick to their own personal descriptors throughout each study and to refrain from discussing their terms with others.

To analyse the observer scores generated, they applied a multivariate statistical technique called Generalised Procrustes Analysis (GPA). This technique does not depend on the use of fixed variables and it permits calculation of the degree of agreement between observers and identification of the commonly perceived dimensions of behavioural expression underlying the observers' separate assessments (for further experimental and statistical details of this methodology, see Wemelsfelder *et al.*, 2000, 2001).

Over the years, they have carried out numerous trials, mostly involving pigs, but also dairy and beef cattle, sheep and poultry. In all of these studies, they claim to demonstrate significant agreement between observers in the interpretation of the animals' behavioural expressions, regardless of these observers' professional background. Individual observers could also repeat their assessments with high levels of accuracy (Wemelsfelder *et al.*, 2001; Rousing and Wemelsfelder, 2006). Significantly, the behaviour shown in the videos used at the FCP trials was frequently also analysed quantitatively using conventional ethograms. Wemelsfelder reported good and meaningful correlations between qualitative and quantitative assessments of behaviour

(Wemelsfelder *et al.*, 2001; Rousing and Wemelsfelder, 2006) and qualitative assessments were also shown to correlate well with physiological measures such as heart rate and heart rate variability (Wemelsfelder, 2007).

They acknowledge that such an approach is often viewed with suspicion because behavioural science traditionally mistrusts expressive terminologies, fearing that these may be based on rather anthropomorphic projections rather than offering reliable assessment of an animal's actual behavioural state (e.g. Kennedy, 1992; Keeley, 2004). However, they argue that in various trials, the assessments of independent observers are consistent (within and between observers) and argue that the results generated do indeed genuinely reflect the animal's subjective experience of its environment and capacity to cope. Wemelsfelder argues that 'the consistent close correlation between observers' qualitative assessments and quantitative measures of behaviour and physiology is important in demonstrating that these assessments have biological validity, and are not just unreliable "subjective" perceptions'.

The approach has also been trialled using observations of animals in groups rather than those observed individually, and demonstrates that observers could reliably judge the expressive quality of larger groups of animals (Wemelsfelder and Farish, 2002).

A way forward in the assessment of animal welfare

Despite these assurances, many people still remain uncomfortable with such qualitative assessments and prefer a focus on more objective, quantitative methods. In addition, it is generally accepted that the most valid assessment of animal welfare is obtained from assessment methods that combine indicators from behavioural observation, alongside indicators of physical condition/health. To be robust such indicators should, according to Rousing *et al.* (2001), should (i) have a firm basis in scientific knowledge and understanding; (ii) have the ability to express development over time; (iii) be measurable within a realistic time frame and (iv) be relevant as a decision support system for the manager.

Clearly many of the physiological measures for welfare status explored in preceding sections are not easily applicable to live animals (requiring invasive sampling) – and many require quite sophisticated (research) methodologies. However, if we accept our general premise that welfare is defined by an ability to adapt and respond to environmental challenge in an appropriate way – (and that thus both positive and negative welfare states are a function of the actual adaptive capacities of the individual animal and the opportunity it has to express those responses) – then our assessment of welfare must remain primarily based on observation of the physiological condition and behavioural responses shown by individual animals over time. If we truly believe that an individual animal's welfare status is defined in terms of its own self-perception of its well-being, then we must rely in major part on behavioural cues to advise us of whether or not that individual (or others in its group) are showing behaviours that are directed towards altering/improving their current status – and whether or not those behavioural responses are likely to be successful.

While accepting that the time required for prolonged observation of any individual animal in farm or laboratory is likely to be impractical and uneconomic for regular use during routine assessments – and that opportunities for such prolonged observation may not even be possible for wild or free-ranging animals – we would still advocate the use of behavioural cues where possible, including the identification of clearly appropriate and adaptive behaviour (or lack of appropriate response in given circumstances) or the expression of appropriate appetitive behaviour – where an animal can be seen to be searching for the appropriate resources to address some perceived deficiency (seeking shelter from wind or adverse weather conditions, etc.).

Such observation should also reveal lack of expected behaviour (even when opportunity to express that behaviour may be present) and also any clearly inappropriate or atypical behaviours (such as an individual separated at some distance from an obvious social group; an animal clearly being rejected or shunned by others within the group; an animal in poor condition suffering continual displacement, etc.). We deliberately seek indicators that are applicable in any context of animal interactions (thus transcend artificial divisions between assessment of welfare status in laboratory animals, in farm livestock, companion animals, etc.) and indeed, given increasing awareness of human responsibility to have concern for the welfare even of free-ranging wildlife affected by human activities (page 52, 53), would seek to develop indicators that could be applied in the same way to fully wild or free-ranging animals.

We would propose that, based on consideration of the animal's adaptive capacity and ability to respond to external challenges, an effective assessment of welfare status could be based on indicators as summarised in Table 8.1. Note that the indicators suggested are presented only in very general terms and more precise definition of behaviours falling into each 'category' would need to be developed in the case of each animal species considered. While such observations may be practicable for individuals or groups that we may study over prolonged periods of time (and may well be applicable to closely managed animals including closely managed wildlife), such an approach is clearly not likely to be feasible in application where encounters with individuals or groups are typically occasional, fleeting and at a considerable distance. Here, inevitably, we must base assessments primarily at the group level and these will be biased in favour of physical condition scores or rather coarse behavioural indicators; such measures will largely be applied at group rather than individual level. This may well be appropriate given that effective management measures (mostly 'non-specific measures', *sensu* Swart, 2005; see Chapter 7) can only be targeted at the group or population level.

But where decisions are made that aim to optimise the welfare of the group as a whole, we should acknowledge, as in Chapter 5, that the welfare status of the group may be optimised while the (apparent) welfare states of its individual members may vary over a considerable range. In interpretation and analysis of such data, therefore, we should recognise that the adaptive capacity of a group or population is unlikely to be well represented simply by the mean welfare status of its individuals because it is to be expected that they will show considerable variation in apparent and actual welfare status. As a consequence, we suggest that it is extremes in (the distribution of) individual welfare status that might be more significant for assessing the status of a group rather than the mean status of its members (and see again page 72).

In effect, the 'mean' value of a given parameter will not necessarily differ between two sets of data that consist of either relatively similar measurements or of measurements that may be distributed over quite a wide range. What *will* differ, however, is the variation of such a set of data and it is that variance which may well be more significant in assessing welfare at the group or population level. The adaptive capacity of the group as a whole is likely to become compromised only if the distribution of the group members shifts towards both extremes (indicating persistent social instability within the group), or to the negative extreme only.

That is not to say we should ignore the mean value totally, since it is clear that a shift of the mean value towards a negative welfare status (of *all* group members) will also indicate a potential welfare problem. Thus, we advocate consideration of both mean and variance of whatever parameters might be employed to assess welfare status.

With such proviso, we may then extend behavioural measures suggested in Table 8.1 in relation to the assessment of individual welfare to develop parallel indicators that may be applied more readily at the level of the group or wider population.

Table 8.1 *Indicators of welfare at the individual and group level, respectively. Assessment should not be based on any single indicator but should attempt to integrate information from as many of these indicators as possible.*

Assessment of Animal Welfare

Based on the animals' adaptive capacities	Individual level		Group level	
	Positive indicators	Negative indicators	Positive indicators	Negative indicators
The animal(s) should be free adequately to react to hunger/thirst.	Appetitive and successful foraging behaviour Normal activity pattern Appropriate body condition	Unsuccessful foraging behaviour Lethargy Inappropriate body condition	Appetitive and successful foraging behaviour and activity pattern as a group Normal variation of body condition	Unsuccessful foraging as a group; successful foraging only in minority of group members (extreme variation within group)
The animal(s) should be free adequately to react to climate conditions.	Seeking and finding shelter Appropriate condition of pelage Appropriate modulation of body condition during seasons	Not finding shelter Bad fur/pelage condition Body condition worse than can be expected in relation to season	Seeking and finding shelter for all group members Appropriate modulation of variation in pelage and body condition during seasons	Not finding shelter or finding insufficient shelter for the group Fur and body condition bad throughout the group or in extreme variation
The animal(s) should be free adequately to react to physical injury or disease.	Seeking and finding rest and shelter Functional immune system (e.g. appropriate wound healing/lack of scouring)	Inability to seek and find shelter Infection/inappropriate wound healing; persistent scouring	Functional immune system (e.g. appropriate wound healing; lack of scouring)	Signs of infection across (parts of) the group (e.g. persistent scouring)

The animal(s) should be free to express its full non-social behavioural repertoire.	Adequate behavioural responses to non-social circumstances/challenges (covering both avoidance and approach behaviours)	Persistent behavioural inhibition, lethargy, context-inadequate behaviour	Adequate behavioural responses to non-social circumstances/challenges that involve the group as a whole (covering both avoidance and approach behaviours)	Behavioural responses that do not involve the whole group
The animal(s) should be free to respond adequately to social interactions.	Adequate behavioural responses to social interactions (covering both socio-positive and socio-negative behaviours)	Persistently being bullied (in social species); social isolation	Social stability within the group (as displayed by adequate socio-positive and socio-negative behaviours)	Social instability; splitting up in sub-groups
The animal(s) should be free to experience the full spectrum of emotional states and respond to those states adequately.	Executing anxiety-related behaviour and stress-responses, as well as play-or other pleasure-related behaviour in appropriate context	Inadequate emotional responses (lethargy, hyperreactivity); absence of adequate emotional responses (e.g. lack of anxiety)	Displaying anxiety-related behaviour and stress-responses, as well as play-or other pleasure-related behaviour at the group-level and in appropriate context	Absence of pleasure-related behaviour; inadequate emotional responses (lethargy, hyperreactivity) at the level of the group

9

ALTERNATIVES TO ANIMAL EXPERIMENTATION: REFINE, REDUCE AND REPLACE!

JAN VAN DER VALK

Introduction

In chapter 7 we discussed how some degree of animal suffering may be tolerated by society, even when it is possible to achieve further reductions, if (perhaps in an agricultural production setting) it would be wholly uneconomic to do so. A certain level of suffering may also be 'acceptable' where laboratory animals are, for instance, used in pre-clinical trials of veterinary medicines or newly developed drugs for human medicine. However, use of animals in research would only be acceptable to society if there were no replacement methods available and if tests were properly contrived to use only the minimum number of test subjects and minimise the necessary suffering.

Towards this end, a number of initiatives have been developed to "Refine, Reduce and Replace' experimentation. European legislation under Directive 2010/63/EU acknowledges 'the importance of the protection and welfare of animals used for scientific purposes at international level' (recital 3, Directive 2010/63/EU) and basically boils down to the 'Not-unless' principle. Animal experiments are not allowed unless the scientific results sought cannot be obtained otherwise. Where experimentation is permitted, every project and procedure requiring animals, whether scientific or educational, will have to undergo a rigorous project evaluation before being authorised. Project evaluation should ensure that there are indeed no available methods that do not require use of animals to obtain the results sought, and should balance the likely harm to the animal against the expected benefits of the project. It should also include ethical considerations, 'both on moral and scientific grounds, to ensure that each use of an animal is carefully evaluated as to the scientific or educational validity, usefulness and relevance of the expected result of that use' (recital 39, Directive 2010/63/EU).

When no replacement models appear to be available to obtain the results sought, assurance of the proper consideration of welfare of experimental animals is regarded as extremely important for several reasons. For purely ethical reasons we should try to avoid inflicting any suffering on animals during such experiments as far as is possible. But also for scientific reasons: animals that are compromised in their welfare do not show 'normal' behaviour or physiological responses, will respond to treatments differently and won't serve as a relevant model even for its own species, let alone for humans.

3Rs – Replacement, Reduction, Refinement

A crucial part of project evaluation before experimentation may be authorised is whether the 3Rs have been considered and, where possible, applied. The 3Rs refer to consideration of every practical means to replace, reduce and refine animal experiments. Sometimes the 3Rs are still referred to as 'alternatives to animal experiments'. Confusingly, some regard alternatives to animal experiments only in terms of replacement. Furthermore, the term 'alternatives' implies a choice between (two) mutually exclusive possibilities, which is, under the current legislation, not the case here: when non-animal methods are available to obtain the results sought, animal experiments are prohibited! For that reason, it is preferred to use the term '3Rs-methods'.

The 3Rs were originally described by their 'godfathers' Bill Russell and Rex Burch in their book *The Principles of Humane Experimental Technique* (Burch, 1959). They defined the 3Rs as follows:

- Replacement means the substitution for conscious living higher animals of insentient material.
- Reduction means reduction in the number of animals used to obtain information of given amount and precision.
- Refinement means any decrease in the incidence or severity of inhumane procedures applied to those animals which still have to be used.

The essence of the 3Rs has more recently been elegantly captured in a song by Bill Russell (1996):

In laboratory buildings from the attic to the basement,
there are people working busily to bring about replacement
In the field of education we expect to see a tutor,
using virtual reality provided by computer.
And instead of living animals with feelings and sensations,
we're already doing better with in vitro preparations.
When the rules of validation are as mad as any hatter,
just a little rationality will remedy the matter...

In the future when experimenters have all received instruction,
we can really make the best of methods of reduction.
To avoid the waste of animals the very first condition,
is to get designer programs from a special statistician.
With an optimal environment for rearing and for testing,
and the breeds precisely labelled from the list of Michael Festing.
Note the age and sex and status whether thin or rather fatter,
so that others can repeat it right and in every little matter....

Biomedical experiments whatever the assignment,
are invaluably better for the maximum refinement.
Not the slightest of distresses are inflicted with impunity,
they're upset by biochemistry and sabotage immunity.
To minimise distresses you must understand the ballet,
that is animal behaviour studied ethologic halley.
This particularly rapid unintelligible patter,
isn't generally heard of and if it is it doesn't matter....

In the next few pages we offer some examples of 3Rs models. Since there are many models in the biomedical field, no attempt will be made to offer comprehensive coverage; we will restrict ourselves simply to some illustrative examples.

Replacement

It is obvious that experiments on humans will give the best results where experiments are being undertaken for human rather than possibly veterinary application. However, ethical, practical and legal objections still limit the use of humans as subjects in research. Nonetheless, we do see increasing applications where humans or human tissues can be used. Processes in the human body can be studied using modern imaging techniques like NMR, MRI, PET and CT scans, enabling the real time and non-invasive study of kinetics and effects of drugs in the body, organ functions like brain structures and activities, and its diseases. Microdosing too is a relatively new technique that studies the behaviour of compounds at extremely low non-pharmacologically active doses. The test compounds are radioactively labelled, enabling the study of the biokinetics of the compound and its possible metabolism. Since these studies can be directly performed in humans, it saves research animals and avoids animal to human translation errors (Svendsen *et al.*, 2015)

However, there are many types of explorations which cannot ethically be performed with human subjects, at least before some initial element of pre-screening. It is here that experiments may be licensed on animals and where increasingly there has been intense effort to search for 3Rs methods. Some tests can be carried out on isolated tissues, on cell cultures or more complex organoids derived from stem cell cultures. There is also a developing technology for organs-on-a-chip. These organs-on-a-chip are essentially plastic and/or gel-based microdevices about the size of a USB stick in which one or several chambers containing cell cultures are connected with small channels making it possible for them to interact.

But we should recognise that all the replacement models that have been and are currently being designed – even these sophisticated organs-on-a-chip technologies – mimic only part of the activities of a complex living organism, without necessarily taking account of the interaction of the system under scrutiny with other organ systems in the way which would occur in an *in vivo* system. Thus, whatever the technologies employed, at some point more information is required than may be available from replacement models, whether these may be studies in biokinetics to find out whether a compound, drug or chemical is able to reach the target organ at the required dose, or studies in biotransformation to explore whether the drug or chemical may perhaps be metabolised into compounds that have different effects from the ones intended. Crucial for every model is its evaluation or validation: a study of its reproducibility and exploration of which aspects of the system being simulated are correctly mimicked by the model and where it fails.

Often, in effect, replacement models replace only one part of a study, or are used in an initial screening: thereby sometimes avoiding the need for further animal studies, but certainly often reducing the number of animal studies and leading to better designed experiments. In fundamental research, non-animal models are used to get better insight in certain processes. But we must recognise that, depending on the outcome, results may often still have to be verified in a complete organism. Only when an animal experiment is replaced by a non-animal model in experiments that routinely take place, for instance for testing of compounds, pharmaceuticals or vaccines, or in education do we have a real replacement.

In vitro models/ex vivo models

In vitro models, cell and tissue culture methods, are popular ways to study processes at the cellular and organ level. The source of the cells and organs are primary cells, cell lines and stem cells.

Primary cells are collected from donors, either animals or, preferably, humans. Unless the cells are obtained from animals that are killed for another purpose, methods that use cells from research animals cannot be regarded as full replacement models; since the animals are bred specifically for the donation of their cells and/or organs, and also because more experiments can be performed with the tissue of any one animal, they should actually be regarded as reduction methods. Some therefore name these models 'relative replacement' models. Taking a step further towards complete replacement, however, it is possible in many instances to use tissues from slaughtered animals that are not suitable for human or animal consumption, to help towards an independence from animals specifically bred to provide these tissues. Examples might be bones for the study of bone formation, joints for the study of, for instance, arthritis, and eyes for studying eye irritation – a particularly nice example is the application of the eyes of slaughtered chickens for assessing the irritation potential of different compounds (Schutte *et al.*, 2009).

There are inevitably problems with such models, however. Cell lines grow infinitely in culture and are a homogenous and continuous source of cells. But because these cells proliferate continuously, which most cells *in vivo* do not, they may not express all features of the cells they originated from. Further, although the cells and organs cultured are collected from a complete organism, they may not fully display their *in vivo* functions in culture because the interactions which would occur with other organ systems within the complete organism are missing. The cells and tissues grow and are maintained in an artificial culture medium and may therefore behave differently. In addition, most cells require the use of animal-derived elements in the medium, like bovine serum albumin and foetal bovine serum, to proliferate and maintain themselves. The composition of these supplements is largely unknown, and may differ between batches, thereby introducing an issue of reproducibility (van der Valk *et al.*, 2004).

Stem cells can also be a good source for study (Bauman, 2017). By exposing these cells to specific growth factors, they can differentiate into specific cells of the type required for study or, under the right conditions, develop into a complex organoid containing the different cell types of the tissue under study (see below). But, at this moment, stem cells still have their limitations because they don't always seem to mature fully into the desired cells.

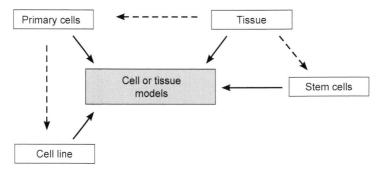

Figure 9.1 *Sources for cell and tissue models.*

To produce antigen-specific antibodies (also named monoclonal antibodies), it was the practice in the past to inject animals intraperitoneally with the antigen concerned in order to harvest the antibodies produced: a procedure which commonly caused substantial suffering. Nowadays, we use hybridoma cells in culture bottles and hollow fibre systems to produce monoclonal antibodies without the use of animals (Goding, 1980). Hybridoma cells are the result of fusing antibody-producing cells with a tumour (myeloma cells). Plant cells and algae are also being used as antibody producers (Yusibov *et al.*, 2016). Modern phage-display technologies have improved the production of antibody fragments (Carmen and Jermutus, 2002).

Cell cultures, particularly those with cell lines, generally contain only one cell type and are limited in their applications because information on, for instance, biokinetics and metabolism are missed. To mimic the more complex build-up of organs, organotypic cultures have been developed. These are either complete organs, parts of an organ or cultured organs from embryonic or adult mammalian stem cells. The latter are also called organoids (e.g. Clevers, 2016). Different organoids like gut, kidney and brain have been developed. These allow the study of the development of those organs but also their pathologies and diseases.

The aim of an organotypic culture is to maintain or represent the structural and functional relationships between the cells and tissues of the organ that is being studied. The disadvantage of these systems, due to their compact 3D structure, is that nutrients have difficulty in reaching the inner cells, and removal of waste material is problematic. The lifespan is therefore limited. Nice examples are the different 3D-skin models that have been developed to study skin irritation: Episkin, Skinethic and Epiderm

(Schäfer-Korting *et al.*, 2008). These models are based on human skin from which different skin cells were isolated, multiplied (proliferated) in culture and reconstructed. These models play an important role in assessing skin irritation of cosmetic ingredients for which, nowadays, no animal studies are allowed in many regions of the world.

New developments, like the organ-on-a-microchip, might overcome some of the challenges mentioned above. Organ-on-a-chip models, as mentioned earlier, can be plastic or gel-based microdevices in which one or several cell-containing chambers are connected with small channels. The use of microfluidic technologies makes it possible to supply nutrients to the tissues in the different chambers and transport away of waste material. These models can give a better representation of biological activities, with a well-controlled microenvironment in which functional mechanical properties and biochemical functionalities better mimic (human) organ functions. The first organ-on-a-chip was a functioning lung model developed by the Wyss Institute (Lind *et al.*, 2016; see also https://wyss.harvard.edu/technology/human-organs-on-chips/). This lung model was developed with human alveolar epithelial cells on one side of a porous membrane and human pulmonary microvascular endothelial cells on the other side, in which the peristaltic movement of the lung was also mimicked. The model allows for a continuous stream of a blood substitute, which can also contain immune cells. Current developments of this chip technology seek to have different organ systems on one or several chips, interconnected by common vascular channels. Such developments ultimately offer the hope of developing a complete 'human-body-on-a-chip'.

All these models could be used some day to identify new therapeutic agents, systemic responses to new drugs, and research into interactions at the cell and organ level and between organs. This might give more insights into how diseases develop and what possible therapies can be developed (Bhatia and Ingber, 2014). *In vitro* methods are cost-effective, more reliable and reproducible compared to most animal experiments. In particular, when the *in vitro* model is based on human cells or tissues, it will facilitate translation to the human situation. For an overview of validated *in vitro* methods, see http://alttox.org/mapp/table-of-validated-and-accepted-alternative-methods/

In vivo models using other animal species

To avoid or replace a need for experiments on higher vertebrates, experiments may instead be carried out on lower organisms deemed to have lower sensitivity and thus posing less of an ethical issue. Lower species like invertebrates (worms, insects, crustacea) are commonly substituted as models for higher animals in studies of developmental biology, neurobiology, genetics, reproduction and mutagenesis. The use of the fruitfly *Drosophila melanogaster* in the study of genetic mechanisms is well-known (e.g. Gilbert, 2008), while sea urchins have been used extensively in studies of reproduction (Graillet *et al.*, 1993) and bacteria are used in the standard Ames test to study the mutagenic properties of different compounds (Ames *et al.*, 1973). Blood from the horseshoe crab (*Limulus*) is used to detect pyrogens in injection fluids, a test which was previously carried out in rabbits. This Limulus Amebocyte Lysate (LAL) test, forms a lysate very sensitively when it comes into contact with an endotoxin (Levin and Bang, 1964) and is now widely used in the pharmaceutical and medical devices industries.

The use of invertebrates in this way causes fewer ethical issues, although research has shown that some species do show behaviour that in complexity equals that of vertebrates and many commentators have concerns, from a welfare perspective, that some species may have quite advanced levels of sentience and awareness. For that reason, cephalopods, such as octopus and squid, are now regarded as fully sentient beings and studies using these animals now require a project evaluation in Europe, in the same way as proposed procedures using higher vertebrates.

Computer modelling

Based on information already available, computer programs (also named *in silico* models) have been developed, for instance, to simulate diseases and predict how new drugs will act in the body. Virtual human organs and metabolism programs help to understand processes inside the human body. Quantitative Structure-Activity Relationships (QSARs) are computer models that can estimate whether or not a substance is hazardous by comparing its properties with existing chemicals and information on human and animal biology. One example is the DEREK system (Deductive Estimation of Risk from Existing Knowledge; Benfenati and Gini, 1997). In addition, in the sphere of estimating ADME outcomes of compounds (Absorption, Distribution, Metabolism, Excretion), several computer programs have been developed based on existing data from animal studies. ADME characteristics of chemicals facilitate interpretation of hazard data and making subsequent risk assessments (Selick *et al.*, 2002; see also http://alttox.org/replacing-animals-in-absorption-distribution-metabolism-excretion-studies/).

Integrated Approaches to Testing and Assessment

Integrated Approaches to Testing and Assessment (IATA) combine and intergrate data from several complementary stand-alone tests, such as those described earlier, and also newly developed (*in vitro*) methods, so that the combined results give a (more) complete picture of the fate of a compound in the body and improve the toxicological decision making (Worth and Patlewicz, 2016). These methods are mostly used for risk or safety assessment of compounds. They are often also applied in combination with the weight of evidence approach where the outcome of the different models is weighed and included in the decision-making process. Well-designed integrated testing methods have shown to drastically reduce animal experiments for a specific purpose up to 85% (Verbost *et al.*, 2007). But although these methods may reduce the number of experimental animals or even replace animal studies, their implementation, in particular in a regulatory context, faces many challenges, especially with respect to validation.

Reduction and refinement

Experimental design

Experimental design is crucial for reducing the number of animals required in experimental studies and for increasing the relevance of those studies. For any experimental study involving animals it is of paramount importance that the results will give useful (significant) results. If too few animals are used in the experiment, results will not reach significance; on the other hand, experiments using too large a number of animals are wasteful and may cause some level of suffering to more animals than necessary.

One of the main issues here is the variation in response shown to any given challenge by different individuals – even those of established genetic lines that have been inbred and maintained as homogeneous as possible for many generations. We have already introduced the problems arising from differences between individuals in both behavioural and physiological responses to a given challenge, in terms of the development of assessment methods for determining welfare status (see Chapter 4). This same variation in response, even to a constant and standardised situation, also poses problems for experimental testing. A large variation in responses confounds the result and requires that large numbers of individuals must be used in any given test in order to reduce statistical variance to an acceptable level. Work currently in progress (M. van der Groot, *in progress*) seeks ways of pre-screening potential trial subjects with simple non-invasive tests in order to select from the potential candidates those which show the lowest variation in response to standard procedures. In this way, it should be possible in the future to reduce the number of animals required in any test by pre-selection of those likely to show a more uniform response – allowing greater statistical significance of results with fewer, pre-screened individuals.

To improve the reproducibility of results of different trials, it is also crucial that publications describe carefully the methods used in housing and rearing experimental animals. Recommendations, like the ARRIVE guidelines, facilitate the publication of methods in more detail and improve reproducibility of results between successive trials (Kilkenny *et al.*, 2010). This, however, introduces another major way in which the number of experiments undertaken might be reduced. Those planning to undertake any experimental procedure should first devote adequate time to screening the available literature to ensure that the work they are proposing has not already been undertaken elsewhere – or that the experimental approach proposed has not already be shown by other workers to be inappropriate or ineffective. It is essential that new trials involving animal subjects shall not simply be repeating work already carried out and in effect 'a rediscovery of the wheel'.

Sharing animal tissues and information

Within the same context, nowadays there is an increasing focus on 'sharing' animals and information. The EU Directive requires that Member States should facilitate the sharing of organs and tissues of animals that are killed (art. 18, Dir. 2010/63/EU). Scientists who need certain organs for their studies do not necessarily have to breed their own animals. They might also use control animals from other studies, surplus animals derived from breeding programmes, or use animals from which other organs are taken for other studies. This way of sharing tissues from 'ethically sourced' animals requires good communication and a matching system in each institute, and also between institutes. Matching systems are being developed and increasingly used (see, for instance, Animatch (https://www.animatch.eu/), Searchbreast (https://searchbreast.org/) and ShARM (http://www.sharmuk.org/)).

Specialised websites may also contribute to refinement and reduction by collecting information on a single subject from different sources and making this available in a well-designed, accessible and searchable way. These websites facilitate the search for specific information and reduce the risk of missing information. The Humane Endpoints website (https://www.humane-endpoints.info/), for example, provides information about humane endpoints and educates how to recognise these in rodents. By applying the humane endpoints, unnecessarily suffering of experimental animals is prevented. Another example is the Experimental Design Assistant (EDA). EDA is an online tool to guide researchers through the design of their experiments, helping to ensure that they use the minimum number of animals consistent with their scientific objectives, methods to reduce subjective bias and appropriate statistical analysis.

Data sharing is a relative new attempt to reduce animal experimentation. The ideal should be that not only data from successful studies should be made available, but also those of unsuccessful studies (which currently are often not reported in scientific literature, thereby increasing the chance of repetition of these unsuccessful experiments by others). Also, combining data from different studies gives opportunities to answer new questions without animal experiments. Although data sharing still faces many challenges, to reduce testing on non-human primates and accelerate biomedical progress, researchers agreed that data sharing is the way of the future (Brouillette, 2017).

Animal care, education and training

Where no obvious replacement methods are available and animals must still be used in experimental procedures, it is crucial that these animals receive good care and every consideration is given to their welfare and to minimise suffering. This requires establishment of good practice within Animal Houses and breeding facilities as well as in actual experimentation. Much of this is dependent not only on the standard of housing and other aspects of direct care, but also on the competence, and thus education and training, of the persons involved. Those caring for animals in experimental facilities need to be educated, trained and certified to provide the best care for animals before, during and after experimental conditions. The same holds for the persons that perform experimental procedures on animals. They must be skilled in the procedures, but also have knowledge of, and know when and how to apply analgesia, anaesthesia and euthanasia to avoid unnecessary suffering of the animals. Good education and training form the basis for skilled and competent persons in animal technologies and contribute to refinement.

In such education and training, the use of 3Rs methods should be emphasised if only to drive home the point that the use of animals in experiments is not always necessary and that the 3Rs methods we have been discussing are available and are valid models. When individuals have to be trained in specific animal skills, training should start on artificial models. Handling rats and mice can be simply exercised with fluffy toys. For training of more invasive techniques, for instance injection techniques for mice and rats, there are now excellent models that enable handling, restraining and practising the different injection techniques (see, for instance, the Koken rat,[8] a life size representation of a 9-week-old male rat with anatomically correct skull, larynx, windpipe, stomach and tail with veins). Since operation skills also include elements like anaesthesia, pulsating blood flow, observing the animal (heart rate, temperature, etc.), at a certain moment skills training has to be completed with real animals, but with these training models, the use of animal for these training purposes can be decreased up to 90% (Remie, 2001).

Discussion

In fundamental research, new methods and models are often developed because these give better opportunities to study processes that are difficult or impossible to study in a complete and fully functional organism. They are developed because of new knowledge and technologies, and often because of multidisciplinary approaches. The organ-on-a-chip technologies, for instance, are dependent on both the input from biologists, specialists in micro-fluid dynamics and often material experts. These new developments may become popular in the short term, stimulating knowledge and further development as we have also seen with organoid systems. Although not intentionally developed to reduce animal experimentation, several models will eventually lead to reduction of animal experiments.

On the other hand, specific methods are being developed to replace experimental animals in standardised routine studies carried out to fixed protocols, such as those often used for regulatory purposes to study the safety of chemicals or in the study of the efficacy/potency and safety of drugs and vaccines. However, given the inertia of the system and the difficulties in altering the legal requirements of evidence deemed acceptable for registration of compounds for commercial use, it is often very challenging to get these replacement models used and implemented. Any new model proposed should be sensitive but, preferably, also selective and must in addition be fully validated. Validation studies are particularly focused on reliability (reproducibility) and relevance. Current safety practices are designed in a way that optimal human safety was guaranteed (low number of false negatives), but may have many false positives, leading to failure of potentially useful products, drugs and vaccines. Although the validation process is currently under discussion, in general the new model is tested with many reference chemicals and compared to the results of animal experiments. Ironically, however, most animal models have never been validated!

---

8 KOKEN RAT LM-046A: Koken Co. Ltd., Tokyo.

When a model or method is regarded as validated, i.e. it correctly assesses the effects of many different compounds, it has to be taken up in regulatory guidelines. Regulatory authorities have to ensure the safe use of different compounds and products. They will be reluctant to move away from (animal-based) methods that have been shown to be reasonably reliable with respect to safety and liability and start using methods with which they have no experience. In addition, harmonisation all over the world is crucial, otherwise the internationally operating producer/manufacturer is faced with performing required animal studies for areas where the new method is not (yet) accepted and will be hesitant to implement animal-free methods. Finally, the new methods have to be used by the users: testing labs and contract research organisations.

In brief, the introduction of the new methods is still a time-consuming process, including validation and acceptance by regulatory authorities and internationally operating organisations (World Health Organization (WHO), Organisation for Economic Co-operation and Development (OECD)). It takes a lot of time, money and effort to eventually reduce or replace animal experiments in a regulatory context. Efforts should be made to shorten this process in order to speed up acceptance and use of animal-free methods.

Non-animal methods do not always fully replace animal tests but function as an adjunct to, or prior to, animal experiments. Based on the results, certain compounds can be selected for further testing in animals. Drugs with desired effects and devoid of undesired (toxic) effects can therefore be selected for further studies. Several methods are now widely used by the pharmaceutical industry in so-called high-throughput screening methods to identify possible new drug candidates, but the chemicals industry also uses this high-throughput system to detect undesired effects of chemicals.

Conclusion

Whether animal-free biomedical research will ever be accomplished has yet to be seen. Recently, documents have been published that explore possibilities to obtain animal-free research before specific deadlines (NCad: Netherlands National Committee for the protection of animals used for scientific purposes, 2016). The conclusions are challenging and promising, and animal-free research might be possible in the near future in specific areas such as protocol studies for regulatory purposes.

In other areas, studies have still to be performed with animals. Focus here should be on methods that reduce animal numbers and approaches that affect the welfare of the experimental animals the least. For now, human safety and health are regarded as so important that animal research is still accepted under controlled conditions.

Despite the development of many different animal-free models, these have their limitations because they are still 'models'. Developments that might reduce some of the limitations by adding complexity to the models, like human-on-a-chip using organoids, and integrated testing methods, are still in their infancy and have many hurdles to overcome. Sensitive imaging and detection techniques might facilitate earlier and safer studies in humans. Still, in studies where the animal is the target species, for veterinary purposes, we will never get rid of animal experiments. Focus there will be on better and reproducible experiments in which the lowest number of animals are used to give relevant data and where animal welfare is minimally compromised.

Although we see many new 3Rs methods being developed and in use, in practice we do not see a corresponding reduction of animal experiments. That has to do in some part with an improvement in speed of experiments. While replacement methods may reduce the number of animals needed for a particular purpose, because these methods also speed up experiments, the reduction of animals is not always shown in animal statistics that are based on time as reference. But we also see new developments in animal modification techniques, like CRISPR/cas9, a revolutionary gene editing technology that enables efficient and precise genomic modifications in a wide variety of organisms and tissues (Doudna and Charpentier, 2014). These new developments are quickly picked up by the scientific community to explore their possibilities in different research areas and therefore often lead to an increased animal usage which, at the same time, counteracts the effects of 3Rs models. Furthermore, we have to realise that in order to develop an animal-free model, knowledge about the processes we would like to study should be available and that information can only be obtained or, in a later stage, be confirmed in a living complex organism. Even the development of 3Rs models themselves still requires animal studies. In addition, many studies for animal purposes, like development of drugs and vaccines in the veterinary practice, are often developed with the target animal and will have to be tested in the target animal. Full replacement of experimental animals, therefore, cannot be expected in the short run, if ever.

10

Concluding Remarks

A proper concern for the welfare of animals emerges as a moral obligation from a number of different philosophical schools of thought – at least in relation to animals under human control. Many legal instruments reflect this presumption, with legislation in many Westernised countries providing a duty of care for animals. In practice, however, this extends in most cases simply to a responsibility not to cause 'unnecessary suffering' and applies only where the animals concerned are deemed to be 'under the control of man, on a permanent or temporary basis'.

Responsibilities towards wild or free-ranging animals are a less universal consequence, with such obligations arising primarily from those schools of thought which accord animals moral status as sentient beings or simply as being of intrinsic value. Even those philosophies which require concern for sentient beings do not extend consideration to all animals – and there is indubitably a grey area as to what defines sentience and which taxonomic groups are considered sentient or not. Furthermore, we might debate what constitutes being under human control – whether at a philosophical or purely legislative level. As in Chapter 7, we may argue that at least within a Westernised context, all animals (whatever their status as kept or non-kept) are to some degree influenced by human activity – whether to a greater or lesser extent, and whether deliberate or incidental. Whether or not directly managed by man, wild animals suffer significant impacts from our activities in loss of habitat to agriculture, forestry or urban sprawl, loss of connectivity of habitats through the proliferation of road systems or other transport infra-structure, impacts on food, etc. and animals are regularly killed and injured on our roads. Many wildlife populations are also directly managed by humans (for control, recreation or conservation) by culling or through use of other, non-lethal, control measures, such as translocation, imposed contraception, and so on. Thus, even in terms of those philosophical considerations which base responsibilities for animals on their relationship with humans, a consideration of the effects of human activity on welfare might be expected even for apparently free-ranging wild animals, as well as those more obviously under human control.

But it is clear that even this level of commitment is primarily a product of Westernised philosophies. Attitudes to animals vary tremendously within that Western world and certainly when one extends consideration more widely around the globe. Ethical views and standards vary tremendously from culture to culture and with ethical tradition, so that there is no universality of the status of animals in society or what our obligations to them may be, closely managed or otherwise. It is important to understand those ethical and cultural differences because, as we emphasise in Chapters 2 and 7, whatever may be our understanding of the biology of animal welfare and whatever approaches we may take to assess the welfare status of animals, this welfare status is, *per se*, neither morally judged good or bad; it simply reports a factual condition. Decisions as to whether a given welfare status is considered a welfare *issue* and whether or not there is an obligation for humans to act in a way to try and improve that status (whether a legal or a moral obligation) can only be taken in relation to the prevailing cultural norms and traditions of human society, which vary significantly from culture to culture, adding to the complexities of variation arising from different ethical schools of thought. (By the same token, whatever our own judgement, we should be wary of imposing attitudes derived from our own society's norms on others with a different cultural or philosophical tradition).

In this complex interplay between science and societal values, the biology is probably the simplest part – if nonetheless contentious! Although early commentators variously defined welfare as primarily a function of health and well-being, or all to do with emotions, it is now generally agreed that welfare describes an internal state of an individual, as experienced by that individual. As Webster (2012) writes: 'There is now broad agreement amongst academics and real people that the welfare of a sentient animal is defined by how well it feels; how well it is able to cope with the physical and emotional challenges to which it is exposed'. In such context we ourselves would emphasise that welfare status is also a function of the animal's capacity to respond appropriately with a suite of physiological and behavioural adaptation to stresses imposed by its environment and thus a dynamic process related to the animal's own (individual) adaptive capacity and the extent to which it is free to exercise its ability to respond to any given challenge.

This idea that animals have generally evolved adaptations to their environment, optimising the ability to adapt to changes within that environment through the expression of a variety of physiological and/or behavioural responses, was first applied within a welfare context nearly four decades ago (see Chapter 3) and it seems to us remarkable that it has taken so long for the concept to become more generally adopted; it is still by no means universally reflected in the wider literature.

Adherence to such a dynamic view of animal welfare, with welfare status related to the animal's own adaptive capacity has, however, a number of implications: we cannot know for certain the adaptive capacity of all individual animals we may encounter because these may vary significantly between individuals, in relation to the physiological and behavioural repertoire they may have inherited or acquired through individual experience. As we have discussed in Chapter 4, individuals may also vary in 'personality' and coping style, as well as in the extent to which they may satisfice rather than optimise. This implies that there can be no hard and fast checklists for the assessment of welfare based on environmental conditions (Chapter 8), but that assessments must be based on observations of the behaviour of the animal or animals concerned as they attempt to adjust to the given challenge with which they may be confronted. A negative state of 'welfare' is therefore perceptible via reactions that are aimed at changing the existing situation. A positive state of 'welfare' is perceptible via reactions aimed at keeping the existing situation as it is. Human determination of the successful adaptive functioning of an individual or a group of individuals is only as good as the observer's perception of the signals that those animals emit.

It becomes clear that if we accept that welfare is defined by an *ability to adapt and respond to environmental challenge in an appropriate way* – and that both positive and negative welfare states are a function of the actual adaptive capacities of the individual animal and the opportunity it has to express those responses – then our assessment of welfare status must be primarily based on detailed observation of the physiological condition and behavioural responses. Such assessment cannot therefore be instantaneous, but implies observation over some given period of time.

But how should we define an appropriate timescale? Recognition of a need to relate welfare to adaptive capacity also implies that short periods of negative experience may not constitute a compromise to welfare status since, as long as such negative states are not prolonged, they may actually be necessary to trigger the appropriate adaptive response from the animal's own adaptive repertoire. This therefore implies that observation should be continued until such period has elapsed that the animal has itself had ample opportunity to deploy the appropriate responses from its adaptive repertoire before we can make a judgement that a challenge is beyond its adaptive capacities to resolve. However, there needs to be a recognition that we cannot wait too long if it is clear that the animal is unable to perform the appropriate adaptive response (either because it does not exist within its personal behavioural or physiological repertoire or because it is in some way physically prevented from doing so); nor should the timescale be extended to a point where the animal is experiencing a negative state over a prolonged period (suffering).

Throughout this book we have emphasised that considerations of welfare should not be restricted to those simply concerned with the avoidance of actual suffering – some societies and ethical traditions may also accept an implicit responsibility for active facilitation of positive welfare-states. Positive emotional states may include pleasure, comfort, contentment, curiosity and playfulness, which suggests that regular observation of such 'positive' behaviour-types might argue for the individual(s) concerned being in a status that it perceives as positive. But a recent advance notice for the 2017 conference of the Universities' Federation for Animal Welfare posed the legitimate questions: 'Is a permanent state of positive welfare possible, or do animals reset their emotional state so that attempts to achieve positive welfare are doomed to failure as the animal habituates to a better than adequate environment?' and 'What happens when those experiences preferred by an animal have a long-term negative impact on health?'

For now, at least, whatever our wider interpretation of our moral responsibilities, legal obligations remain effectively restricted to avoidance of suffering (and generally an obligation restricted to closely managed animals – primarily farm livestock and laboratory animals). Within the European Union, for example, all Member States are required to provide legal protection from unnecessary suffering for managed animals, but the EU has as yet no remit to provide for consideration of animal welfare *per se*. Germany has comprehensive animal welfare laws and animal welfare is included in the German constitution, but laws operate in a similar manner to most other European countries regarding unnecessary suffering and only place a duty on people who care for an animal or have it in their possession, although countries such as the Netherlands and Norway have empowered a more widespread responsibility for welfare (see Chapter 7).

The science and management of animal welfare still has a long way to go – and legislation still further to progress if it is to catch up with current understanding.

How might such considerations impact on veterinarians and other welfare professionals? The minimalist approach to welfare implied by legislation is generally regarded, even by an informed public, as inadequate. While much of this legislation is based on the Brambellian tradition of the Five Freedoms, it is apparent that the Five Freedoms were never originally conceived as more than a minimum safeguard to protect the welfare of farm animals over the years and were never intended to define welfare *per se*, or to be necessary and sufficient to ensure positive welfare. We believe that in addition to any responsibility to address issues of negative welfare veterinarians, as the frontline champions of animal welfare, should seek actively to promote positive welfare in all contexts of engagement. Many indeed do so.

But while the veterinarian's basic professional responsibility towards animals is clearly independent from the context of animal use, it is obvious that their decisions in daily practice and those of other welfare professionals must be guided or constrained by a context-dependent legal provision. This might argue a responsibility within the profession to challenge this compartmentalisation of legislation and to stimulate the discussion as to whether current legal frames appropriately reflect recent developments in scientific and ethical thinking and are thus appropriate in promoting a more consistent approach to welfare management (Yeates, 2009). Such an approach might also embrace a more dynamic view of welfare, where welfare is intimately related to the animal's own capacity to respond appropriately to any given challenge. Yeates (2012) has noted that 'A more dynamic concept of welfare may lead to a less interfering approach to our patients and a greater appreciation that conservative options are not always just cheaper "second-best" options', but also suggested that 'our paternalism is probably here to stay, and so too are the Five Freedoms. Both are probably too entrenched in the veterinary consciousness to be successfully altered, however insightful the underlying logic'. We might hope that this is not the case.

In conclusion, to structure discussion at a workshop we were holding, Putman and Ohl were asked to present a form of personal 'Credo', summarising in a single slide what we considered the most important tenets of modern welfare science.

We responded as follows.

◇ We believe that a concept of welfare based on (the Five Freedoms and) the avoidance of suffering is minimalist and restrictive.

◇ We believe that an animal's welfare lies on some continuum from poor to good and that we should seek to manage animals and their environment not only to avoid negative welfare/suffering, but actively to promote good welfare ('A life worth living').

◇ We believe in a more dynamic view of animal welfare such that welfare is related to an animal's adaptive capacities.

◇ We believe that welfare should therefore not be considered as an instantaneous construct to be assessed at some moment in time. An adaptive response may take some finite period of time; crucially, therefore, our assessment of welfare should not simply consider the status of any individual at a given moment in time, but needs to be integrated over the longer time periods required to execute such change.

◇ We believe that an animal's welfare *status* is in major part a function of its own perception and response to its own condition. That welfare status is, intrinsically, neither morally to be judged good or bad; it simply is.

◇ We believe that self-assessment/perception of a given status shows significant inter-individual variation and that there is also significant variation in adaptive capacity and coping strategy.

◇ We believe that in consequence, the (objectively determined) welfare status of all members of any group of animals may appear to vary over a considerable range, yet all members perceive their own welfare state as optimal – or at least satisfactory.

◇ We believe that such conclusion implies that purely objective functional scales for measuring the welfare status of individual animals can have little validity since, even under identical conditions, the actual welfare status of different individuals may very widely.

◊ We believe that, because of this expectation of high variation in apparent welfare, that in attempting to safeguard satisfactory welfare in populations or groups of individuals, we must insert into protocols some minimum threshold value below which no individual should be allowed to fall, instead of (or in addition to) simply determining some average welfare status to be achieved.

◊ We believe that a responsibility for safeguarding animal welfare is universal and independent of context and thus a responsibility which extends to free-ranging wildlife as well as more closely managed animals directly under human care, but that in some circumstances intervention may be constrained or overruled by practicalities or perceived risk of unintended consequences.

References

Adam, S. and Kriesi, H. (2007) The network approach. In: P.A. Sabatier (ed.), *Theories of the Policy Process*. Westview Press, Cambridge, pp. 129–154.

Allen, K., Shykoff, B.E. and Izzo, J.L. (2001) Pet ownership, but not ACE inhibitor therapy, blunts home blood pressure responses to mental stress. *Hypertension* **38**, 815–820.

Ames, B.N., Lee, F.D. and Durtson, W.E. (1973) An improved bacterial test system for the detection and classification of mutagens and carcinogens. *PNAS* 70(3), 782–786. Available at https://dx.doi.org/10.1073%2Fpnas.70.3.782

Appleby, M.C. and Sandøe, P. (2002) Philosophical debate on the nature of well-being: implications for animal welfare. *Animal Welfare* **11**, 283–294.

Armstrong, S.J. and Botzler, R.G. (2003) General introduction. In: S.J. Armstrong and R.G. Botzler (eds.), *The Animal Ethics Reader*. Routledge, London, pp. 1–11.

Ascione, F.R. and Shapiro, K. (2009) People and animals, kindness and cruelty: research directions and policy implications. *Journal of Social Issues* **65**(3), 569–587.

Ascione, F.R., Weber, C.V., Thompson, T.M., Heath, J., Maruyama, M. and Hayashi, K. (2007) Battered pets and domestic violence: animal abuse reported by women experiencing intimate violence and by non-abused women. *Violence Against Women* **13**(4), 354 –373.

Ausems, E. (2006) The Council of Europe and animal welfare. In: Council of Europe (ed.), *Animal Welfare*. Council of Europe, Ethical Eye Series, Strasbourg, pp. 233–253.

Baldry, A.C. (2003) Animal abuse and exposure to interparental violence in Italian youth. *Journal of Interpersonal Violence* **18**, 258. doi: 10.1177/0886260502250081.

Barker, S.B. and Dawson, K.S. (1998) The effects of Animal Assisted Therapy on patients' anxiety, fear, and depression before ECT. *The Journal of ECT* **19**(1), 38–44.

Barnes, J.E., Boat, B.W., Putman, F.W., Dates, H.F. and Mahlman, A.R. (2006) Ownership of high-risk ('vicious') dogs as a marker for deviant behaviors: implications for risk assessment. *Journal of Interpersonal Violence* **21**(12), 1616–1634.

Barnett, J.L. and Hemsworth, P.H. (1990) The validity of physiological and behavioural measures of animal welfare. *Applied Animal Behaviour Science* **25**, 177–187.

Bartholomew, K. (1991) Avoidance of intimacy: an attachment perspective. *Journal of Social and Personal Relationships* **7**, 147–178.

Bartolomé, E., Sánchez, M.J., Molina, A., Schaefer, A.L., Cervantes, I. and Valera, M. (2013) Using eye temperature and heart rate for stress assessment in young horses competing in jumping competitions and its possible influence on sport performance. *Animal* **7**, 2044–2053.

Bateson, P. and Bradshaw, E.L. (1997) Physiological effects of hunting red deer (*Cervus elaphus*). *Proceedings of the Royal Society of London B* 264, 1707–1714.

Bauman, K. (2017) Stem cells: a key to totipotency. *Nature Reviews Molecular Cell Biology* doi: 10.1038/nrm.2017.9.

Becker, N., Dillitzer, N., Sauter-Louis, C. and Kienzle, E. (2012) Fütterung von Hunden und Katzen in Deutschland (Feeding of dogs and cats in Germany). *Tierärtz Praxis Kleintiere* 2012(6), 391–397.

Beekman, V., Kaiser, M., Sandoe, P., Brom, F., Millar, K. and Skorupinski, B. (2006) *The development of ethical bio-technology assessment tools for agriculture and food Production.* Final Report. Available at http://edepot.wur.nl/2608 and www.ethicaltools.info/

Benfenati, E. and Gini, G. (1997) Computational predictive programs (expert systems) in toxicology. *Toxicology* **119**(3), 213–225.

Bentham, J. (1789; reprinted 1907) *Introduction to the Principles of Morals and Legislation.* Clarendon Press, Oxford.

Berger, J. (1986) *Wild Horses of the Great Basin.* University Press, Chicago, IL.

Bergvall, U.A., Schäpers, A., Kjellander, P. and Weiss, A. (2011) Personality and foraging decisions in fallow deer, *Dama dama. Animal Behaviour* **81**, 101–112.

Berman, W.H. and Sperling, M.B. (1996) The structure and function of attachment. In: M.B. Sperling and W.H. Berman (eds.), *Attachment in Adulthood: Clinical and Developmental Perspectives.* Guilford Press, New York, NY, pp. 3–28.

Bhatia, S.N. and Ingber, D.E. (2014) Microfluidic organs-on-chips. *Nature Biotechnology* 32, 760–772. doi: 10.1038/nbt.2989.

Blumstein, T. (2010) Conservation and animal welfare issues arising from forestry practices. *Animal Welfare* **19**, 151–157.

Boddice, R. (2009) *A History of Attitudes and Behaviours toward Animals in Eighteenth- and Nineteenth-Century Britain: Anthropocentrism and the Emergence of Animals.* Mellen, Lewiston, NY.

Boissy, A., Manteuffel, G., Jensen, M.B., Moe, R.O., Spruijt, B., Keeling, L.J., Winckler, C, Forkman, B., Dimitrov, I., Langbein, J., Bakken, M., Veissier, I. and Aubert, A. (2007) Assessment of positive emotions in animals to improve their welfare. *Physiology and Behaviour* **92**, 375–397.

Bracke, M.B.M. (2007) Animal-based parameters are no panacea for on-farm monitoring of animal welfare. *Animal Welfare* **16**, 229–231.

Bracke, M.B.M. and Hopster, H. (2006) Assessing the importance of natural behaviour for animal welfare. *Journal of Agricultural and Environmental Ethics* **19**, 77–89.

Bradshaw, J.S.W. and Casey, R.A. (2007) Anthropomorphism and anthropocentrism as influences in the quality of life of companion animals. *Animal Welfare* (Suppl.) **16**, 149–154.

Braithwaite, V. (2010) *Do Fish Feel Pain?* Oxford University Press, Oxford.

Brambell Committee (1965) *Report of the Technical Committee to Enquire into the Welfare of Animals Kept under Intensive Livestock Husbandry Systems.* Cmnd. 2836, December 3 1965. Her Majesty's Stationery Office, London.

Broom, D.M. (1991) Animal welfare: concepts and measurement. *Journal of Animal Sciences* **69**(10), 4167–4175.

Broom, D.M. (1988) The scientific assessment of animal welfare. *Applied Animal Behaviour Science* **20**, 5–19.

Broom, D.M. (1998) Welfare, stress, and the evolution of feelings. *Advances in the Study of Behavior* **27**, 371–403.

Broom, D.M. (2006) Adaptation. *Tierärztliche Wochenschrift* **119**, 1–6.

Broom, D.M. (2007) Quality of life means welfare: how is it related to other concepts and assessed? *Animal Welfare* **16**(Suppl. 1), 45–53.

Broom, D.M. (2010) Cognitive ability and awareness in domestic animals and decisions about obligations to animals. *Applied Animal Behaviour Science* **126**, 1–11.

Brosnan, K. (2011) Do the evolutionary origins of our moral beliefs undermine moral knowledge? *Biology and Philosophy* **26**, 51–64.

Brouillette, M. (2017) To treat primates more humanely: transparency. *Scientific American* February 2017. Available at https://www.scientificamerican.com/article/to-treat-primates-more-humanely-transparency/

Burton, D. (1992) *The Effects of Parasitic Nematode Infection on Body Condition of New Forest Ponies.* PhD thesis, University of Southampton, UK.

Cabanac, M. (1971) Physiological role of pleasure. *Science* **173**, 1103–1107.

Cabanac, M. (1979) Sensory pleasure. *Quarterly Reviews in Biology* **54**(1), 1–29.

Callicott, J.B. (1980). Animal liberation: a triangular affair. *Environmental Ethics* 2, 311–338.

Carmen, S. and Jermutus, L. (2002) Concepts in antibody phage display. *Briefings in Functional Genomics and Proteomics* **1**, 189–203.

Carpenter, E. (1980) *Animals and Ethics.* Watkins, London.

Chandler, C.K. (2005) *Animal Assisted Therapy in Counseling.* Routledge, New York, NY.

Clevers, H. (2016) Modeling development and disease with organoids. *Cell* 165(7), 1586–1597. http://dx.doi.org/10.1016/j.cell.2016.05.082.

Clutton-Brock, T. (2002) Breeding together: kin selection and mutualism in cooperative vertebrates. *Science* **296**, 69–72.

Cockram, M.S., Shaw, D.J., Milne, E., Bryce, R., McClean, C. and Daniels, M.J. (2011) Comparison of effects of different methods of culling red deer (*Cervus elaphus*) by shooting, on behaviour and post mortem measurements of blood chemistry, muscle glycogen and carcase characteristics *Animal Welfare* **20**, 211–224.

Cohen, N.E., Brom, F.W.A. and Stassen, E.N. (2009) Fundamental moral attitudes to animals and their role in judgment: an empirical model to describe fundamental moral attitudes to animals and their role in judgment on the culling of healthy animals during an animal disease epidemic. *Journal of Agricultural and Environmental Ethics* **22**, 341–359.

Colby, P.M. and Sherman, A. (2002) Attachment styles impact on pet visitation effectiveness. *Anthrozoös* **15**, 150–165.

Cole, K.M., Gawlinski, A., Steers, N. and Kotlerman, J. (2007) Animal-assisted therapy in patients hospitalized with heart failure. *American Journal of Critical Care* **16**, 575–585.

Coleman, W.D., Skogstad, G.D. and Atkinson, M.M. (1996) Paradigm shifts and policy networks: cumulative change in agriculture. *Journal of Public Policy* **16**, 273–301.

Colonius, T.J. and Earley, R.W. (2013) One Welfare: a call to develop a broader framework of thought and action. *Journal of American Veterinary Medical Association* **242**(3), 309–310.

Coppens, C.M., de Boer, S.F. and Koolhaas, J.M. (2010) Coping styles and behavioural flexibility: towards underlying mechanisms. *Philosophical Transactions of the Royal Society* B **365**, 4021–4028.

Costa, M.J.R.P. (2003) Principios de etoligia aplicados ao bem-estar das aves. *Anais da Conferencia Apinco de Ciencia e Tecnologis Avicola*; Campinas, 160–177.

Courcier, E.A., Mellor, D.J., Thomson, R.M. and Yam, P.S. (2011) A cross sectional study of the prevalence and risk factors for owner misperceptions of canine body shape in first opinion practice in Glasgow. *Preventive Veterinary Medicine* **102**, 66–74.

Crofoot, M.C., Rubenstein, D.I., Maiya, A.S. and Berger-Wolf, T.Y. (2011) Aggression, grooming and group-level cooperation in white-faced capuchins (*Cebus capucinus*): insights from social networks. *American Journal of Primatolology* **73**, 821–833.

Crowley, P.H. and Baik, K.H. (2010) Variable valuation and voluntarism under group selection: an evolutionary public goods game. *Journal of Theoretical Biology* **265**, 238–244.

Currie, C.L. (2006) Animal cruelty by children exposed to domestic violence. *Child Abuse and Neglect* **30**(4), 425–435.

Dantzer, R. and Mormede, P. (1983) Stress in farm animals: a need for re-evaluation. *Journal of Animal Science* **57**, 6–18.

Dantzer, R., Mormede, P. and Henry, J.P. (1983) Significance of physiological criteria in assessing animal welfare. In: D. Smidt (ed.), *Indicators Relevant to Farm Animal Welfare*. Martinus Nijhoff, pp. 29–37.

Dawkins, M.S. (1990) From an animal's point of view: motivation, fitness and animal welfare. *Behavioural and Brain Sciences* **13**, 1–31.

Dawkins, M.S. (1998) Evolution and animal welfare. *Quarterly Review of Biology* **73**, 305–328.

Dawkins, M.S. (2003) Behaviour as a tool in the assessment of animal welfare. *Zoology* **106**, 383–387.

Dawkins, M.S. (2008) The science of animal suffering. *Ethology* **114**, 937–945.

Dawkins, R. (1980) Good strategy or evolutionary stable strategy? In: G.W. Barlow and S. Silverberg (eds.), *Sociobiology: Beyond Nature/Nurture.* Westview Press, Boulder, CO, pp. 331–367.

de Fontenay, E. (2006) Do animals have rights? In: Council of Europe (ed.), *Animal Welfare.* Council of Europe, Ethical Eye Series, Strasbourg, pp. 29–38.

deKloet, E.R., Joels, M. and Holsboer, F. (2008a) Stress and the brain: from adaptation to disease. *Nature Reviews Neuroscience* **6**, 463–475.

deKloet, E.R., Karst, H and Joels, M. (2008b) Corticosteroid hormones in the central stress response: quick and slow. *Frontiers in Neuroendocrinology* **29**, 268–272.

Dingemanse, N.J., Kazem, A.J.N., Réale, D. and Wright, J. (2009) Behavioural reaction norms: animal personality meets individual plasticity. *Trends in Ecology and Evolution* **25**(2), 81–89.

Donaldson, S. and Kymlicka, W. (2011) *Zoopolis: A Political Theory of Animal Rights.* Oxford University Press, Oxford.

Doudna, J.A. and Charpentier, E. (2014) The new frontier of genome engineering with CRISPR-Cas9. *Science*, 346, 1258096-3–1258096-9. doi: 10.1126/science.1258096.

Dunbar, M.R., Johnson, S.R., Rhyan, J.C. and McCollum, M. (2009) Use of infrared thermography to detect thermographic changes in mule deer (*Odocoileus hemionus*) experimentally infected with foot-and-mouth disease. *Journal of Zoo and Wildlife Medicine* **40**, 296–301.

Duncan, I.J.H. (1993) Welfare is to do with what animals feel. *Journal of Agricultural and Environmental Ethics* **6**(Suppl. 2), 8–14.

Duncan, I.J.H. (1996) Animal welfare defined in terms of feelings. *Acta Agriculturae Scandinavica (Section A: Animal Science)* **27**(Suppl.), 29–35.

Duncan, I.J.H. (2005) Science-based assessment of animal welfare: farm animals. *Revue Scientifique et Technique – Office International des epizooties* **24**(2), 483–492.

Duncan, I.J.H. and Dawkins, M.S. (1983) The problem of assessing 'well-being' and 'suffering' in farm animals. In: D. Smidt (ed.), *Indicators Relevant to Farm Animal Welfare.* Martinus Nijhoff, The Hague, pp. 13–24.

Duncan, I.J.H. and Fraser, D. (1997) Understanding animal welfare. In: M. Appleby and B. Hughes (eds.), *Animal Welfare.* CAB International, Wallingford, UK, pp. 19–31.

Duncan, I.J.H. and Petherick, J.C. (1991) The implication of cognitive processes for animal welfare. *Journal of Animal Science* **69**, 5017–5022.

Edgar, J.L., Nicol, C.J., Pugh, C.A. and Paul, E.S. (2013) Surface temperature changes in response to handling in domestic chickens. *Physiology and Behavior* **119**, 195–200.

Eilam, D., Zor, R. and Szechtman, H. (2006) Rituals, stereotypy and compulsive behavior in animals and humans. *Neuroscience and Biobehavioural Reviews* **30**, 456–471.

El-Alayli, A., Lystad, A.L. Webb, S.R., Hollingsworth, S.L. and Ciolli, J.L. (2006) Reigning cats and dogs: a pet-enhancement bias and its link to pet attachment, pet-self similarity, self-enhancement, and well-being. *Basic and Applied Social Psychology* **28**, 131–143.

Endenburg, N. (1991) *Animals as Companions. Demographic, Motivational and Ethical Aspects of Companion Animal Ownership.* PhD thesis, University of Utrecht.

Endenburg, N. (1995) The attachment of people to companion animals. Anthrozoös **8**, 83–89.

Enders-Slegers, J.M.P. and Janssens, M.A. (2009) Cirkel van geweld: verbanden tussen dierenmishandeling en huiselijk geweld (Circle of violence: links between animal abuse and domestic violence). Wageningen University Library: library. wur.nl.

Endo, Y. and Shiraki, K. (2000) Behavior and body temperature in rats following chronic foot shock or psychological stress exposure. *Physiology and Behavior* **71**, 263–268.

European Union (EU) (2016) Attitudes of Europeans towards animal welfare. *Special Eurobarometer* 442. doi: 10.2875/884639.

Fagen, R. and Fagen, J.M. (1996) Individual distinctiveness in brown bears, *Ursus arctos* L. *Ethology* **102**, 212–226.

Farm Animal Welfare Council (FAWC) (2009) *Farm Animal Welfare in Great Britain, Past, Present and Future.* Farm Animal Welfare Council, London, pp. 243–254.

Feaver, J., Mendl, M. and Bateson, P. (1986) A method for rating the individual distinctiveness of domestic cats. *Animal Behaviour* **34**, 1016–1025.

Ferrari, C., Pasquaretta, C., Carere, C., Cavallone, E., von Hardenberg, A. and Réale, D. (2013) Testing for the presence of coping styles in a wild mammal. *Animal Behaviour* **85**, 1385–1396.

Fogany, P. and Target, M. (1997) Attachment and reflective function: their role in self-organisation. *Development and Psychopathology* **9**, 679–700.

Fraser, A.F. and Broom, D.M. (1990) *Farm Animal Behaviour and Welfare.* Baillière Tindall, London.

Fraser, D. (1993) Assessing animal well-being: common sense, uncommon science. In: *Food Animal Well-being.* Conference Proceedings and Deliberations. Purdue University Office of Agricultural Research Programs, West Lafayette, IN. pp. 37–54.

Fraser, D. (2003) Assessing animal welfare at the farm and group level: the interplay of science and values. *Animal Welfare* **12**, 433–443.

Fraser, D. (2008) Animal welfare, values, and mandated science. In: D. Fraser (ed.), *Understanding Animal Welfare: The Science in its Cultural Context*. Wiley-Blackwell, Oxford, UK, pp. 260–274.

Fraser, D. and Duncan, I.J.H. (1998) Pleasures, pains and animal welfare: toward a natural history of affect. *Animal Welfare* **7**, 383–396.

Fraser, D., Weary, D.M., Pajor, E.A. and Milligan, B.N. (1997) A scientific conception of animal welfare that reflects public values. *Animal Welfare* **6**, 187–205.

Friedmann, E., Katcher, A., Lynch, J. and Thomas, S. (1980) Animal companions and one-year survival of patients after discharge from a coronary care unit. *Public Health Reports* **95**, 307–312.

Gallagher, B., Allen, M. and Jones, B. (2008) Animal abuse and intimate partner violence: researching the link and its significance in Ireland, a veterinary perspective. *Irish Veterinary Journal* **61**, 658.

Gamborg, C., Palmer, C. and Sandøe, P. (2012) Ethics of wildlife management and conservation: what should we try to protect? *Nature Education Knowledge* **3**, 8.

Garcia Pinillos, R., Appleby, M.C., Manteca, X., Scott-Park, F., Smith, C. and Velarde, A. (2016) One Welfare – a platform for improving human and animal welfare. *Veterinary Record* **179**, 412–413.

German, A.J. (2006) The growing problem of obesity in dogs and cats. *Journal of Nutrition* 136(Suppl.), 1940S–6S.

Gewirtz, J.L. and Boyd, E.F. (1997) The infant conditions, the mother. In Th. Alloway, P. Pliner and L. Krames (eds.), *Advances in the Study of Communication and Affect, Attachment Behavior.* Plenum Press, New York, NY, 109–143.

Gilbert, L.I. (2008) *Drosophila* is an inclusive model for human diseases, growth and development. *Molecular and Cellular Endocrinology* 293(1–2), 25–31. http://dx.doi.org/10.1016/j.mce.2008.02.009.

Gill, E.L. (1988) *Factors Affecting Body Condition of New Forest Ponies.* PhD thesis, University of Southampton, UK.

Gill, E.L. (1991) *Factors Affecting Body condition in Free-Ranging Ponies.* Technical Report, Royal Society for the Prevention of Cruelty to Animals (RSPCA), Horsham, UK.

Goding, J.W. (1980) Antibody production by hybridomas. *Journal of Immunological Methods* 39(4), 285–308. http://dx.doi.org/10.1016/0022-1759(80)90230-6.

Gosling, S.D. (1998) Personality dimensions in spotted hyenas (*Crocuta crocuta*). *Journal of Comparative Psychology* **112**, 107–118.

Gosling, S.D. (2001) From mice to men: what can we learn about personality from animal research? *Psychological Bulletin* **127**, 45–86.

Graillet, C., Pagano, G. and Girard, J.-P. (1993) Stage-specific effect of teratogens on sea urchin embryogenesis. *Teratogenesis, Carcinogenesis, and Mutagenesis* **13**, 1–14.

Hamilton, W. (1964a) The genetical evolution of social behaviour: I. *Journal of Theoretical Biology* **7**, 1–16.

Hamilton, W. (1964b) The genetical evolution of social behaviour: II. *Journal of Theoretical Biology* **7**, 17–52.

Handlin, L., Hydbring-Sandberg, E., Nilsson, A., Ejdebäck, M., Jansson, A. and Uvnäs-Moberg, K. (2011) Short-term interaction between dogs and their owners – effects on oxytocin, cortisol, insulin and heart rate – an exploratory study. *Anthrozoös* **24**, 301–316.

Harding, S. (2014) *Unleashed: The Phenomena of Status Dogs and Weapon Dogs*. Policy Press, Bristol.

Havener, L., Gentes, L., Thaler, B., Megel, M.E., Baum, M.M., Driscoll, F.A., Beiraghi, S. and Ograwl, N. (2001) The effect of companion animals on distress in children undergoing dental procedures. *Comprehensive Pediatric Nursing* **24**, 137–152.

Hayashida, S., Oka, T., Mera, T. and Tsuji, S. (2010) Repeated social defeat stress induces chronic hyperthermia in rats. *Physiology and Behavior* **101**, 124–31.

Haynes, R.P. (2011) Competing conceptions of animal welfare and their ethical implications for the treatment of non-human animals. *Acta Biotheoretica* **59**, 105–120.

Held, S.D.E. and Spinka, M. (2011) Animal play and animal welfare. *Animal Behaviour* **81**, 891–899.

Herrenkohl, T.I., Sousa, C., Tajima, E.A., Herrenkohl, R.C. and Moylan, C.A. (2008) Intersection of child abuse and children's exposure to domestic violence. *Trauma, Violence and Abuse* 9(2), 84–99. doi: 10.1177/1524838008314797.

Herzog, H. (2011) *Some we love, some we hate, some we eat. Why it's so hard to think straight about animals.* HarperCollins, New York, NY.

Hofer, H. and East, M. (1998) Biological conservation and stress. In: A.P. Moller, M. Milinski and P.J.B. Slater (eds.), *Advances in the Study of Behavior*. Academic Press, San Diego, CA, pp. 405–525.

Hothersall, B. and Casey, R. (2012) Undesired behavior in horses: a review of their development, prevention, management and association with welfare. *Equine Veterinary Education* **24**, 479–485.

Insel, T. (1988) Obsessive-compulsive disorder: new models. *Psychopharmacological Bulletin* **24**, 365–369.

ICMO (International Commission on Management of the Oostvaardersplassen) (2006) *Reconciling Nature and Human Interests*. Report of the International Committee on the Management of Large Herbivores in the Oostvaardersplassen. The Hague/Wageningen, Netherlands. Wageningen UR – Wing rapport 018.

ICMO2 (International Commission on Management of the Oostvaardersplassen) (2010) *Natural Processes, Animal Welfare, Moral Aspects and Management of the Oostvaardersplassen*. Report of the second International Commission on

Management of the Oostvaardersplassen. The Hague/Wageningen, Netherlands. Wing rapport 039.

Johnsen, P.F., Johannesson, T. and Sandøe, P. (2001) Assessment of farm animal welfare at herd level: many goals, many methods. *Acta Agriculturae Scandinavica A* Suppl. **30**, 26–33.

Johnson, S.R., Rao, S., Hussey, S.B., Morley, P.S. and Traub-Dargatz, J.L. (2011) Thermographic eye temperature as an index to body temperature in ponies. *Journal of Equine Veterinary Science* **31**, 63–66.

Jones, A.R. and Price, S. (1990) Can stress in deer be measured? *Deer* **8**(1), 25–27.

Jones, A.R. and Price, S. (1992) Measuring the responses of fallow deer to disturbance. In: R. Brown (ed.), *The Biology of Deer*. Springer Verlag, Berlin and New York, pp. 211–216.

Jordan, B. (2005) Science-based assessment of animal-welfare: wild and captive animals. *Review of Science and Technology* **24**, 515–528.

Jordan, T. and Lem, M. (2014) One health, one welfare: education in practice. Veterinary students' experiences with Community Veterinary Outreach. *Canadian Veterinary Journal* **55**(12), 1203–1206.

Kalof, L. and Taylor, C. (2007) The discourse of dog fighting. *Humanity and Society* **31**, 319–333.

Kanamori, M., Suzuki, M., Yamamoto, K., Kanda, M., Matsui, Y., Kojima, E., Fukawa, H., Sugita, T. and Oshiro, H. (2001) A day care program and evaluation of Animal Assisted Therapy (AAT) for elderly with senile dementia. *American Journal of Alzheimer's Disease and other Dementias* **16**(4), 234–239.

Kay, R.N.B. (1979) Seasonal changes of appetite in deer and sheep. *Agricultural Research Council Research Review*.

Kay, R.N.B. and Staines, B.W. (1981) The nutrition of the red deer (*Cervus elaphus*). *Nutrition Abstracts and Reviews B* **51**, 601–21.

Keeley, B.L. (2004) Anthropomorphism, primatomorphism, mammalomorphism: understanding cross-species comparisons. *Biology and Philosophy* **19**, 521–540.

Kennedy, J.S. (1992) *The New Anthropomorphism*. Cambridge University Press, Cambridge.

Kienzle, E., Bergler, R. and Mandernach, A. (1998) Comparison of the feeding behaviour of the man–animal relationship in owners of normal and obese dogs. *Journal of Nutrition* **128**(12), 2777S–2782S.

Kikusi, T., Winslow, J.T. and Mori, Y. (2006) Social buffering: relief from stress and anxiety. *Philosophical Transactions of the Royal Society B* **361**, 2215–2228.

Kilkenny, C., Browne, W.J., Cuthill, I.C., Emerson, M. and Altman, D.G. (2010) Improving Bioscience Research Reporting: The ARRIVE Guidelines for Reporting Animal Research. *Plos Biology*. Available at http://dx.doi.org/10.1371/journal.pbio.1000412

King, J.E. (1999) Personality and the happiness of the chimpanzee. In: F. Dolins (ed.), *Attitudes to Animals: Views in Animal Welfare*. Cambridge University Press, Cambridge, pp. 1–113.

Kirkwood, J.K., Sainsbury, A.W. and Bennett, P.M. (1994) The welfare of free-living wild animals: methods of assessment. *Animal Welfare* **3**, 257–273.

Knierim, U. and Winckler, C. (2009) On-farm welfare assessment in cattle – validity, reliability and feasibility issues and future perspectives with special regard to the Welfare Quality® approach. *Animal Welfare* **18**, 451–458.

Knierem, U., Carter, C.S., Fraser, D., Gartner, K., Lutgendorf, S.K., Mineka, S., Panksepp, J. and Sachser, N. (2001) Good welfare: improving the quality of life. In: D. M. Brown (ed.) *Coping with Challenge: Welfare in Animals Including Humans*, Dahlem University Press, Berlin, pp. 79–100.

Koolhaas, J.M. (2008) Coping style and immunity in animals: making sense of individual variation. *Brain Behaviour and Immunology* **22**, 662–667.

Koolhaas, J.M., de Boer, S.F., Coppens, C.M. and Buwalda, B. (2010) Neuroendocrinology of coping styles: towards understanding the biology of individual variation. *Frontiers in Neuroendocrinology* **31**, 307–321.

Koolhaas, J.M., Korte, S.M., De Boer, S.F., van der Vegt, B.J., van Reenen, C.G., Hopster, H., de Jong, I.C., Ruis, M.A.W. and Blokhuis, H.J. (1999) Coping styles in animals: current status in behavior and stress-physiology. *Neuroscience and Biobehavioral Reviews* **23**, 925–35.

Korte, S.M., Koolhaas, J.M., Wingfield, J.C. and McEwen, B.S. (2005) The Darwinian concept of stress: benefits of allostasis and costs of allostatic load and the trade-offs in health and disease. *Neuroscience and Biobehavioral Reviews* **29**, 3–38.

Korte, S.M., Olivier, B. and Koolhaas, J.M. (2007) A new animal welfare concept based on allostasis. *Physiology and Behaviour* **92**, 422–428.

Korte, S.M., Prins, J., Vinkers, C.H. and Olivier, B. (2009) On the origin of allostasis and stress-induced pathology in farm animals: celebrating Darwin's legacy. *The Veterinary Journal* **182**, 378–383.

Krebs, J.R. and Davies, N.B. (1993) On selfishness and altruism. In: J.R. Krebs and N.B. Davies (eds.), *Introduction to Behavioural Ecology*. 3rd edition. Blackwell Scientific Publications, Oxford, pp. 265–290.

Krippendorff, K. (1986) *A Dictionary of Cybernetics.* The American Society for Cybernetics, Norfolk, VA.

Lahti, D.C. (2003) Parting with illusions in evolutionary ethics. *Biology and Philosophy* **18**, 639–651.

Le Neindre, P., Guemene, D., Arnould, C., Leterrier, C., Faure, J.M., Prunier, A. and Meunier-Salaun, M.C. (2004) Space environmental design and behaviour: effects of space and environment on animal welfare. *Proceedings of the Global Conference on Animal Welfare*, Paris, 23–25 February 2004, 135–141.

Leader-Williams, N. and Ricketts, C. (1982) Seasonal and sexual patterns of growth and condition of reindeer (*Rangifer tarandus*) introduced into South Georgia. *Oikos* **38**, 27–39.

Levin, J. and Bang, F.B. (1964) The role of endotoxin in the extracellular coagulation of *Limulus* blood. *Bulletin of the Johns Hopkins Hospital* **115**, 265–274.

Lind, J.U., Busbee, T.A, Valentine, A.D., Pasqualini, F.S., Yuan, H., Yadid, M., Park, S.-J, Kotikian, A., Nesmith, A.P., Campbell, P.H., Vlassak, J.J., Lewis, J.A. and Parker, K.K. (2016) Instrumented cardia microphysiological devices via multimaterial three-dimensional printing. *Nature Materials.* doi: 10.1038/nmat4782.

Lotem, A., Fishman, M.A. and Stone, L. (2003) From reciprocity to unconditional altruism through signalling benefits. *Biological Science* **270**, 199–205.

Ludwig, N., Gargano, M., Luzi, F., Carenzi, C. and Verga, M. (2007) Technical note: applicability of infrared thermography as a non invasive measurements of stress in rabbits. *World Rabbit Science* **15**, 199–205.

Lussier, P., Proulx, J. and Leblanc, M. (2005) Criminal propensity, deviant sexual interests and criminal activity of sexual aggressors against women: a comparison of explanatory models. *Criminology* **43**(1), 249–282.

Macfie, J., McEwain, N.L., Houts, M. and Cox, M.J. (2005) Intergenerational transmission of role reversal between parent and child: dyadic and family systems internal working models. *Attachment and Human Development* **7**, 51–65.

Main, D.C.J., Webster, A.J.F and Green, L.E. (2001) Animal welfare assessment in farm assurance schemes. *Acta Agriculturae Scandinavica A* **51**(Suppl. 30), 108–113.

Manning, A. and Serpell, J. (eds.) (1994) *Animals and Human Society: Changing Perspectives.* Routledge, London and New York.

Mathews, F. (2010) Wild animal conservation and welfare in agricultural systems. *Animal Welfare* **19**, 159–170.

Mayes, E. and Duncan, P. (1986) Temporal patterns of feeding behaviour in free-ranging horses. *Behaviour* **96**, 105–129.

Maynard Smith, J. (1982) *Evolution and the Theory of Games.* Cambridge University Press, Cambridge.

McCabe, K.M., Hough, R., Wood, P.A. and Yeh, M. (2001) Childhood and adolescent onset conduct disorder: a test of the developmental taxonomy. *Journal of Abnormal Child Psychology* **29**(4), 305–316.

McCafferty, D.J. (2007) The value of infrared thermography for research on mammals: previous applications and future directions. *Mammal Review* **37**, 207–223.

McEwan, E.K. and Whitehead, P.E. (1970) Seasonal changes in the energy and nitrogen intake in reindeer and caribou. *Canadian Journal of Zoology* **48**, 905–13.

McEwen, B.S., Angulo, J., Cameron, H., Chao, H.M., Daniels, D., Gannon, M.N., Gould, E., Mendelson, S., Sakai, R., Spencer, R. and Woolley, C. (1992) Paradoxical effects of adrenal steroids on the brain: protection versus degeneration. *Biological Psychiatry* **31**, 177–199.

McEwen, L.S., French, C.E., Magruder, N.D., Swift, R.W. and Ingram, R.H. (1957) Nutrient requirements of the white-tailed deer. *Transactions of the 22nd North American Wildlife Conference*, 119–32.

McGreevy, P.D. and Bennett, P.C. (2010) Challenges and paradoxes in the companion-animal-niche. *Animal Welfare* **19**, 11–16.

McGreevy, P.D., Thomson, P.C., Pride, C., Fawcett, A., Grassi, T. and Jones, B. (2005) Prevalence of obesity in dogs examined by Australian veterinary practices and the risk factors involved. *Veterinary Record* **156**, 695–702.

McKeon, P. (2016) *The Thermal Effects of Climate and Social Stress on Captive Fallow Deer* (Dama dama) *as an Indicator of Chronic Stress.* MRes thesis, Ecology and Environmental Biology, University of Glasgow Institute of Biodiversity, Animal Health and Comparative Medicine, UK.

McMillan, F.D., Duffy, D.L., Zawistowski, S.L. and Serpell, J.A. (2015) Behavioral and psychological characteristics of canine victims of abuse. *Journal of Applied Animal Welfare Science* **18**(1), 92–111.

McPhedran, S. (2009) A review of the evidence for associations between empathy, violence, and animal cruelty. *Aggression and Violent Behavior* **14**, 1–4.

Meijboom, F.L.B. and Ohl, F. (2012) Managing nature parks as an ethical challenge: a proposal for a practical protocol to identify fundamental questions. In: T. Potthast, S. Meisch (eds.), *Ethics of Non-Agricultural Land-Management.* Wageningen Academic, Wageningen, The Netherlands, pp. 131–136.

Mellor, D.J. (2004) Comprehensive assessment of harms caused by experimental, teaching and testing procedures on live animals. *Alternatives of Laboratory Animals* **32**(Suppl. 1), 453–457.

Mellor, D.J. (2012) Animal emotions, behaviour and the promotion of positive welfare states. *New Zealand Veterinary Journal* **60**(1), 1–8.

Mellor, D.J. and Bayvel, A.C.D. (2008) New Zealand's inclusive science-based system for setting animal welfare standards. *Journal of Applied Animal Behaviour Science* **113**, 313–329.

Mellor, D.J. and Reid, C.S.W. (1994) Concepts of animal well-being and predicting the impact of procedures on experimental animals. In: R. Baker, G. Jenkin and D.J. Mellor (eds.), *Improving the Well-being of Animals in the Research Environment.* Australian and New Zealand Council for the Care of Animals in Research and Teaching, Glen Osmond, South Australia, pp. 3–18.

Mellor, D.J. and Stafford, K.J. (2001) Integrating practical, regulatory and ethical strategies for enhancing farm animal welfare. *Australian Veterinary Journal* **79**, 762–768.

Mellor, D.J., Patterson-Kane, E. and Stafford, K.J. (2009) *The Science of Animal Welfare.* Wiley-Blackwell, Oxford, Chapters 1–5.

Mench, J.A. (1993) Assessing animal welfare: an overview. *Journal of Agricultural and Environmental Ethics* **6**, 68.

Mendl, M. and Deag, J.M. (1995) How useful are the concepts of alternative strategy and coping strategy in applied studies of social behaviour? *Applied Animal Behaviour Science* **44**, 119–137.

Mendl, M., Burman, O.H. and Paul, E.S. (2010) An integrative and functional framework for the study of animal emotion and mood. *Biological Science* **277**, 2895–2904.

Mepham, B., Kaiser, M., Thorstensen, E., Tomkins, S. and Millar, K. (2006) *Ethical Matrix Manual*, LEI, The Hague. Available at http://edepot.wur. nl/2608

Midgley, M. (1983) *Animals and Why They Matter: A Journey around the Species Barrier*. University of Georgia Press, Athens, GA.

Mitchell, B., McCowan, D. and Nicholson, I.A. (1976) Annual cycles of body weight and condition in Scottish red deer (*Cervus elaphus*). *Journal of Zoology* (London) **180**, 107–127.

Moberg, G.P. (1985) Biological response to stress. In: G.P. Moberg (ed.), *Animal Stress*. American Physiological Society, Bethesda, MD, pp. 27–49.

Moberg, G.P. (1993) Using risk assessment to define domestic animal welfare. *Journal of Agricultural and Environmental Ethics* **6**, 1–7.

Moen, A.N. (1976) Energy conservation by white-tailed deer (*Odocoileus virginianus*) in the winter. *Ecology* **57**, 192–98.

Moen, A.N. (1978) Seasonal changes in heart rate, activity, metabolism and forage intake of white-tailed deer (*Odocoileus virginianus*). *Journal of Wildlife Management* **42**, 715–38.

Moura, D.J., Naas, I.A., Pereira, D.F., Silva, R.B.T.R. and Camargo, G.A. (2006) Animal welfare concepts and strategy for poultry production: a review. *Brazilian Poultry Science* **8**, 137–148.

Myers, J.P. (1983) Conservation of migrating shorebirds: staging areas, geographic bottlenecks, and regional movements. *American Birds* **37**, 23– 25.

Nagengast, S.L., Baun, M.M., Megel, M. and Leibowitz, J.M. (1997) The effects of the presence of a companion animal on physiological arousal and behavioral distress in children during a physical examination. *Journal of Pediatric Nursing* **12**(6), 323–330.

NCad (Netherlands National Committee for the protection of animals used for scientific purposes) (2016) Opinion provided by NCad as to how the Netherlands can become a pioneer in non-animal research. Available at https:// english.ncadierproevenbeleid.nl/latest/news/16/12/15/ncad-opinion-transition-to-non-animal-research

Nesse, M.R. and Ellsworth, P.C. (2009) Evolution, emotions, and emotional disorders. *American Psychology* **64**, 129–139.

Nordenfelt, L. (2011) Health and welfare in animals and humans. *Acta Biotheoretica* **59**, 139–152.

Norton, B. (1995) Caring for nature: a broader look at animal stewardship. In: B. Norton, M. Hutchins, E.F. Stevens and T.L. Maple (eds.), *Ethics on the Ark: Zoos, Animal Welfare and Wildlife Conservation*. Smithsonian Institution Press, Washington, DC, pp. 102–122.

O'Riordan, T. (2004) Environmental science, sustainability, and politics. *Transactions of the Institute of British Geographers* **29**, 234–247.

Ohl, F. and Fuchs, E. (1999) Differential effects of chronic stress on memory processes in the tree shrew. *Cognitive Brain Research* **7**, 379–387.

Ohl, F. and Putman, R.J. (2013a) *The Effect of Levels of Competence on Deer Welfare – Defining and Measuring Welfare*. Report for Scottish Natural Heritage, Inverness.

Ohl, F. and Putman, R.J. (2013b) *Applying Wildlife Welfare Principles to Individual Animals*. Report for Scottish Natural Heritage, Inverness. Available at www. snh.gov.uk/publications-data-and-research/publications/search-the-catalogue/ publication-detail/?id=2072

Ohl, F. and Putman, R.J. (2013c) *Applying Wildlife Welfare Principles at the Population Level*. Report for Scottish Natural Heritage, Inverness. Available at www.snh.gov.uk/publications-data-and-research/publications/search-the-catalogue/publication-detail/?id=2071

Ohl, F. and Putman, R.J. (2014a) Animal welfare at the group level: more than the sum of individual welfare? *Acta Biotheoretica* **62**(1), 35–45.

Ohl, F. and Putman, R.J. (2014b) Welfare issues in the management of wild ungulates. In: R.J. Putman and M. Apollonio (eds.), *Behaviour and Management of European Ungulates*. Whittles Publishing, Caithness, UK, pp. 236–265.

Ohl, F. and Putman, R.J. (2014c) Animal welfare considerations: should context matter? *Journal of Veterinary Science and Research*, **1**, 1–8.

Ohl, F. and van der Staay, F.J. (2012) Animal welfare – at the interface between science and society. *The Veterinary Journal* **192**, 13–19.

Oka, T., Oka, K. and Hori, T. (2001) Mechanisms and mediators of psychological stress-induced rise in core temperature. *Psychosomatic Medicine* **63**, 476–486.

Ortiz, R. and Liporace, J. (2005) Seizure-alert dogs: observations from an inpatient video/EEG unit. *Epilepsy and Behavior* **6**(4), 620–622.

Palmer, C. (2010) *Animal Ethics in Context*. Columbia University Press, New York, NY.

Panksepp, J. (1998) The quest for long-term health and happiness: to play or not to play, that is the question. *Psychological Inquiry* **9**, 56–66.

Panksepp, J. (2011) The basic emotional circuits of mammalian brains: do animals have affective lives? *Neuroscience & Biobehavioural Reviews* **35**(9), 1791–1804.

Parker, G.A. (1984) Evolutionary Stable Strategies. In: J.R. Krebs and N.B. Davies (eds.), *Behavioural Ecology: An Evolutionary Approach*. 2nd edition. Blackwell Scientific, Oxford, pp. 30–61.

Pollock, J.I. (1980) *Behavioural Ecology and Body Condition Changes in New Forest Ponies*. RSPCA Scientific Publications No. 6. Royal Society for the Prevention of Cruelty to Animals (RSPCA), Horsham, UK.

Price, S. and Jones, A.R. (1992) Responses of farmed red deer to being handled. In: R. Brown (ed.), *The Biology of Deer*. Springer-Verlag, Berling and New York, p. 220.

Prins, H.H.T. (1996) *Ecology and Behaviour of the African Buffalo: Social Inequality and Decision Making*. Chapman & Hall, London.

(Raad voor Dierenaangelegenheden) (2010) *Moral Issues and Public Policy on Animals*. Report 2010/02. s'Gravenhage, The Netherlands.

(RDA) Raad voor Dierenaangelegenheden (2012) *Duty of Care: Naturally Considered*. Report 2012/02. s'Gravenhage, The Netherlands.

Ragatz, L., Fremouw, W., Thomas, T. and McCoy, K. (2009) Vicious dogs: the antisocial behaviors and psychological characteristics of owners. *Journal of Forensic Sciences* **54**(3), 699–703.

Rawls, J. (1971, revised edition 1999) *A Theory of Justice.* Belknap Press/Harvard University Press, Cambridge, MA.

Réale, D., Gallant, B.Y., Leblanc, M. and Festa-Bianchet, M. (2000) Consistency of temperament in bighorn ewes and correlates with behaviour and life history. *Animal Behaviour* **60**, 589–597.

Réale, D., Reader, S.M., Sol, D., McDougall, P.T. and Dingemanse, N.J. (2007) Integrating animal temperament within ecology and evolution. *Biological Reviews* **82**, 1–28.

Regan, T. (1983, reprinted 2004) *The Case for Animal Rights.* Routledge, London.

Remie, R. (2001) The PVC-rat and other alternatives in microsurgical training. *Lab Animal* **30**(9), 48.

Rissing, S., Pollock, G., Higgins, M., Hagen, R. and Smith, D. (1989) Foraging specialization without relatedness or dominance among co-founding ant queens. *Nature* **338**, 420–422.

Robertson, I.D. (2003) The association of exercise, diet and other factors with owner-perceived obesity in privately owned dogs from metropolitan Perth WA. *Preventive Veterinary Medicine* **58**, 75–83.

Rohlf, V.I., Toukhsati, S., Coleman, G.J. and Bennett, P.C. (2010) Dog obesity: can dog caregivers' (owners') feeding and exercise intentions and behaviors be predicted from attitudes? *Journal of Applied Animal Welfare Science* **13**, 213–36.

Rollin, B.E. (1981) *Animal Rights and Human Morality.* Prometheus Books, New York, NY.

Rollin, B.E. (2011) Animal pain: what it is and why it matters, *Journal of Ethics* **15**, 425–437.

Rousing, T. and Wemelsfelder, F. (2006) Qualitative assessment of social behaviour of dairy cows housed in loose housing systems. *Applied Animal Behaviour Science* **101**, 40–53.

Rousing, T., Bonde, M. and Sørensen, J.T. (2001) Aggregate welfare indicators into an operational welfare assessment system: a bottom up approach. *Acta Agriculturae Scandinavica A (Animal Science)* **51**, 53–57.

Russell, W. (1996) World Congress on Alternatives. Utrecht, The Netherlands. Available at https://www.facebook.com/JHUCAAT/videos/21450356928/

Russell, W. M. S and Burch, R.L. (1959) (eds.) *The Principles of Humane Experimental Technique.* Methuen, London.

Rutgers, B. and Heeger, R. (1999) Inherent worth and respect for animal integrity. In: M. Dol, M. Fentener van Vlissingen, S. Kasanmoentalib, T. Visser and H. Zwart, (eds.), *Recognizing the Intrinsic Value of Nature.* Van Gorcum, Assen, The Netherlands, pp. 41–53.

Rutherford, K.M.D. (2002) Assessing pain in animals. *Animal Welfare* **11**, 31–53.

Ryder, R. (1970) *Speciesism.* Privately printed leaflet, Oxford.

Ryder, R. (1971) Experiments on Animals. In: S. Godlovitch, R. Godlovitch and J. Harris (eds.), *Animals, Men and Morals*. Grove Press, New York, NY, pp. 41–82.

Sapolsky, R.M. (1993) Endocrinology alfresco: psychoendocrine studies of wild baboons. *Recent Progress in Hormone Research* **48**, 437–468.

Scanlon, T.M. (1998) *What We Owe to Each Other*. Harvard University Press/ Belknap Press, Cambridge, MA.

Scarlett, J.M., Salaman, M.D., New, J.C. and Kass, P.H. (1999) Reasons for relinquishment of companion animals in US animal shelters: selected health and personal issues. *Journal of Applied Animal Welfare Science* **2**, 41–45.

Schäfer-Korting, M., Bock, U., Diembeck, W. and Weimer, M. (2008) The use of reconstructed human epidermis for skin absorption testing: results of the validation study. *ATLA* **36**(2), 161–187.

Schenk, A.M., Ragatz, L.L. and Fremouw, W.J. (2012) Vicious dogs part 2: criminal thinking, callousness, and personality styles of their owners. *Journal of Forensic Sciences* **57**(1), 152–159.

Schopenhauer, A. (1840, reprinted 1976) *Über die Grundlage der Moral*, in Sämtliche Werke, Band I, Stuttgart/Frankfurt am Main.

Schutte, K., Prinsen, M.K., McNamee, P.M. and Roggeband, R. (2009) The isolated chicken eye test as a suitable *in vitro* method for determining the eye irritation potential of household cleaning products. *Regulatory Toxicology and Pharmacology* 54, 272–281. http://dx.doi.org/10.1016/j.yrtph.2009.05.008.

Seger, J. (1989) All for one, one for all: that is our device. *Nature* **338**, 374–375.

Sejian, V., Lakritz, J., Ezeji, T. and Lal, R. (2011) Assessment methods and indicators of animal welfare. *Asian Journal of Animal and Veterinary Advances* **6**, 301–315.

Selick, H.E., Beresford, A.P. and Tarbit, M.H. (2002) The emerging importance of predictive ADME simulation in drug discovery. *Drug Discovery Today* **15**(7), 109–16.

Selye, H. (1950) Stress and the general adaptive syndrome. *British Medical Journal* **1**, 1383–1392.

Serpell, J.A. (2002) Anthropomorphism and anthropomorphic selection – beyond the 'Cute Response'. *Society and Animals* **10**, 437–454.

Shaw, D.S., Bell, R.Q. and Gilliom, M. (2000) A truly early starter model of antisocial behavior revisited. *Clinical Child and Family Psychological Review* **3**, 155–172.

Sheriff, M.J., Dantzer, B., Delehanty, B., Palme, R. and Boonstra, R. (2011) Measuring stress in wildlife: techniques for quantifying glucocorticoids. *Oecologia* **166**, 869–887.

Short, H.L., Newsom, J.D., McCoy, G.L. and Fowler, J.F. (1969) Effects of nutrition and climate on southern deer. *Transactions of the 34th North American Wildlife Conference*, 137–45.

Sih, A. and del Guidice, M. (2012) Linking behavioural syndromes and cognition: a behavioural ecology perspective. *Philosophical Transactions of the Royal Society B* **367**, 2762–2772.

Sih, A., Bell, A. and Johnson, J.C. (2004a) Behavioral syndromes: an ecological and evolutionary overview. *Trends in Ecology and Evolution* 19, 372–378. doi: 10.1016/j.tree.2004.04.009.

Sih, A., Bell, A.M., Johnson, J.C. and Ziemba, R.E. (2004b) Behavioral syndromes: an integrative overview. *Quarterly Review of Biology* **79**, 241–277.

Simon, H.A. (1955) A behavioral model of rational choice. *Quarterly Journal of Economics* **59**, 99–118.

Sinclair, L., Merck, M. and Lockwood, R. (2006) *Forensic Investigation of Animal Cruelty*. Humane Society Press, Washington, DC.

Singer, P. (1975) *Animal Liberation: A New Ethics for our Treatment of Animals*. Random House, New York, NY.

Singer, P. (1989) All animals are equal. In: T. Regan and P. Singer (eds.), *Animal Rights and Human Obligations*. Prentice Hall, Englewood Cliffs, NJ, pp. 148–162.

Singer, P. (1995) *Practical Ethics*. 2nd edition. Cambridge University Press, Cambridge.

Smith, B.R. and Blumstein, D.T. (2008) Fitness consequences of personality: a meta-analysis. *Behavioral Ecology* **19**, 448–455.

Sozmen, B. (2012) Dismissing death: an evaluation of suffering in the wild. Presentation to Conference: 'Animal Welfare: Building Bridges between Science, the Humanities and Ethics'. Abstract on website: www.uu.nl/hum/ mindinganimals. (Click through via programme and then abstracts of parallel sessions).

Stewart, M., Stafford, K.J., Dowling, S.K., Schaefer, A.L. and Webster, J.R. (2008) Eye temperature and heart rate variability of calves disbudded with or without local anaesthetic. *Physiology and Behavior* **93**, 789–97.

Stewart, M., Webster, J.R., Schaefer, A.L., Cook, N.J. and Scott, S.L. (2005) Infrared thermography as a non-invasive tool to study animal welfare. *Animal Welfare* **14**, 319–325.

Stewart, M., Webster, J.R., Verkerk, G.A., Schaefer, A.L., Colyn, J.J. and Stafford, K.J. (2007) Non-invasive measurement of stress in dairy cows using infrared thermography. *Physiology and Behavior* **92**, 520–525.

Svendsen, P., El-Galaly, T.C., Dybkær, K., Bøgsted, M., Laursen, M.B., Schmitz, A., Jensen, P. and Johnsen, H.E. (2016) The application of human phase 0 microdosing trials: a systematic review and perspectives. *Leukemia and Lymphoma* **57**(6), 1281–1290.

Swart, J.A.A. (2005) Care for the wild. Dealing with a pluralistic practice. *Environmental Values* **14**(2), 251–263.

Swart, J.A.A and Keulartz, J. (2011) Wild animals in our backyard. A contextual approach to the intrinsic value of animals. *Acta Biotheoretica* **59**, 185–200.

Taylor, K.D. and Mills, D.S. (2007) Quality of life in companion animals. *Animal Welfare* **16**, 55–65.

Taylor, N.R., Main, D.C., Mendl, M and Edwards, S.A. (2011) Tail-biting: a new perspective. *The Veterinary Journal* **186**(2), 137–47.

Taylor, P. (1986) *Respect for Nature*. Princeton University Press, Princeton, NJ.

Trivers, R.L. (1971) The evolution of reciprocal altruism. *Quarterly Review of Biology* **46**, 35–57.

Valera, M., Bartolomé, E., Sánchez, M.J., Molina, A., Cook, N. and Schaefer, A. (2012) Changes in eye temperature and stress assessment in horses during show jumping competitions. *Journal of Equine Veterinary Science* **32**, 827–830.

van der Burg, W. and van Willigenburg, T. (eds.) (1998) *Reflective Equilibrium. Essays in Honour of Robert Heeger*. Kluwer Academic Publishers, Dordrecht, pp. 89–99.

van der Valk, J., Mellor, D., Brands, R., Fischer, R., Gruber, F., Gstraunthaler, G., Hellebrekers, L., Hyllner, J., Jonker, F.H., Prieto, P., Thalen, M. and Baumans, V. (2004) The humane collection of fetal bovine serum and possibilities for serum-free cell and tissue culture. *Toxicology in Vitro* **18**, 1–12.

Vapnek, J. and Chapman, M. (2010) *Legislative and Regulatory Options for Animal Welfare*. Food and Agriculture Organisation of the United Nations (FAO) Legal Office, Rome.

Veissier, I. and Boissy, A. (2007) Stress and welfare: two complementary concepts that are intrinsically related to the animals' point of view. *Physiology and Behaviour* **92**, 429–433.

Verbost, P.M., van der Valk, J. and Hendriksen, C. (2007) Effects of the introduction of *in vitro* assays on the use of experimental animals in pharmacological research. *ATLA* **35**, 223–228.

Vianna, D.M.L. and Carrive, P. (2005) Changes in cutaneous and body temperature during and after conditioned fear to context in the rat. *The European Journal of Neuroscience* **21**, 2505–2512.

Vis, B. and Hylands, T. (2013) Fat government, thin populace? Is the growth of obesity prevalence lower in more generous welfare states? *International Journal of Social Welfare* **22**(4), 360–73.

Ward, D. (1992) The role of satisficing in foraging theory. *Oikos* **63**, 312–317.

Ward, D. (1993) Foraging theory, like all other fields of science, needs multiple working hypotheses. *Oikos* **67**, 376–378.

Warren, M.A. (1997) *Moral Status*. Oxford University Press, Oxford.

Webster, A.J.F. (1994) *A Cool Eye Towards Eden*. Blackwell Science, London.

Webster, A.J.F. (2011) Zoomorphism and anthropomorphism: fruitful fallacies? *Animal Welfare* **20**, 29–36.

Webster, A.J.F. (2012) Critical control points in the delivery of improved animal welfare. *Animal Welfare* **21**(S1), 117–123.

Weiss, A., King, J.E. and Perkins, L. (2006) Personality and subjective well-being in orangutans (*Pongo pygmaeus* and *Pongo abelii*). *Journal of Personality and Social Psychology* **90**, 501–511.

Wells, D.L. and Hepper, P.G. (2012) The personality of 'aggressive' and 'non-aggressive' dog owners. *Personality and Individual Differences* 53(6), 770–773.

Wemelsfelder, F. (1997) The scientific validity of subjective concepts in models of animal welfare. *Applied Animal Behaviour Science* **53**, 75–88.

Wemelsfelder, F. (2003) Seeing animals whole. *Resurgence* (January/February), 20–21.

Wemelsfelder, F. (2007) How animals communicate quality of life: the qualitative assessment of behaviour. *Animal Welfare* **16**(S), 25–31.

Wemelsfelder, F. and Farish, M. (2002) The qualitative assessment of pig behaviour and welfare in litter groups (Abstract). *Second International Workshop on the Assessment of Animal Welfare at Farm or Group level.* School of Veterinary Science, University of Bristol, UK, p. 27.

Wemelsfelder, F. and Lawrence, A.B. (2001) Qualitative assessment of animal behaviour as an on-farm welfare-monitoring tool. *Acta Agriculturae Scandinavica A*, Suppl. **30**, 21–25.

Wemelsfelder, F., Hunter, E.A., Mendl, M.T. and Lawrence, A.B. (2000) The spontaneous qualitative assessment of behavioural expressions in pigs: first explorations of a novel methodology for integrative animal welfare measurement. *Applied Animal Behaviour Science* **67**, 193–215.

Wemelsfelder, F., Hunter, T., Mendl, M. and Lawrence, A. (2001) Assessing the whole animal: a free choice profiling approach. *Animal Behaviour* **62**, 209–220.

West, S.A., Griffin, A.S. and Gardner, A. (2007) Social semantics: altruism, cooperation, mutualism, strong reciprocity and group selection. *Journal of Evolutionary Biology* **20**, 415–432.

Whaytt, H.R., Main, D.C.J., Green, L.E. and Webster, A.J.F. (2003) Animal-based methods for the assessment of welfare state of dairy cattle, pigs and laying hens: consensus of expert opinion. *Animal Welfare* **12**, 205–217.

Wood, A.J., Cowan, I.McT. and Nordan, H.C. (1962) Periodicity of growth in ungulates as shown by deer of the genus *Odocoileus. Canadian Journal of Zoology* **40**, 596–603.

Worth, A.P. and Patlewicz, G. (2016) Integrated approaches to testing and assessment. In: C. Eskes and M. Whelan (eds.), *Validation of Alternative Methods for Toxicity Testing. Advances in Experimental Medicine and Biology 856.* Springer International Publishing, Cham, Switzerland, pp. 317–342.

Yeates, J.W. (2009) Response and responsibility: an analysis of veterinary ethical conflicts. *The Veterinary Journal* **182**, 3–6.

Yeates, J.W. (2012) How should veterinary surgeons adapt to achieve animal welfare? *The Veterinary Journal* **192**, 6–7.

Yeates, J.W. and Main, D.C.J. (2008) Assessment of positive welfare: a review. *The Veterinary Journal* **175**, 293–300.

Yusibov, V., Kushnir, N. and Streatfield, S.J. (2016) Antibody production in plants and green algae. *Annual Review of Plant Biology* 67, 1–729. doi: 10.1146/annurev-arplant-043015-111812.

Zahavi, A. (1995) Altruism as a handicap: the limitation of kin selection and reciprocity. *Journal of Avian Biology* **26**, 1–3.

Glossary of terms

It remains one of the problems of any new area of enquiry – and the welfare 'area' is no exception – that successive authors tend to adopt words from common parlance and give them quite technical meanings. In order to avoid misunderstandings and potential misinterpretation, we therefore felt it might be useful to offer a glossary of terms to make clear what is meant by each term – by us or by others – when used in this book.

Adaptive capacity

We use the expression 'adaptive capacity' to describe the set of physical and mental abilities with which an animal is able to respond and 'adapt' to its environmental situation and any challenges it may encounter. Many features of this adaptive capacity have been acquired by a species through evolution; others may be developed by individual animals as a result of their own lifetime experience. The species-specific abilities form a basis, which is refined and developed in each individual. The adaptive capacity of an individual is not static; it continues to develop throughout an animal's life. In addition, at any one instant of time it is dependent on the individual's internal state as well as on changing environmental conditions.

In considering the welfare of individuals of more social species, where an animal's environment also includes other members of its social group, it is necessary to re-evaluate the adaptive capacities of an individual as being related to the functioning of a social group as a whole.

The adaptive capacity of a group describes the set of (physical and mental) abilities with which a group of animals is naturally endowed. The species-specific abilities of each group-member form a basis which is refined and developed in each individual as a functional part of the whole. As with that of the individual, the adaptive capacity of a group is not static; it is dependent on the interactive functioning of group members as well as on changing environmental conditions.

Adaptive response

Welfare is in large part a function of an animal's ability to respond appropriately and in some adaptive way to its environmental circumstances. Adaptive responses thus are characterised by behavioural or physiological responses that enable an individual (or group) to react appropriately to both positive and potentially harmful (negative) stimuli (e.g. approach a food resource or avoid a potential danger).

Appropriate adaptive response

There may be considerable variation between individuals in their adaptive capacities (their actual behavioural or physiological abilities to respond adaptively) and in the actual coping strategy adopted. Different coping strategies reflect what some authors describe as different underlying 'personalities' of individual animals, where some may be more active and others more passive in response. As a result, there may be quite substantial variation in the way different individuals may respond to the same stressor and the ways in which they may cope with environmental or social challenge – not simply in relation to differences in the adaptive repertoire available to different individuals, but also in relation to coping strategy adopted. While it may thus be 'appropriate' for an actively coping individual to respond more pro-actively or aggressively to a given challenge, avoidance of that same stimulus might be appropriate for a more passively coping individual.

Depending on internal (e.g. hormonal or developmental) and external changes (e.g. season) an individual may respond differently even to the same stimulus at different times. While such different responses may all be adaptive, a distinct response may be more appropriate at a given juncture depending on prevailing internal and/or external circumstances. For example, foraging behaviour is clearly adaptive; however, during harsh weather conditions it might be more appropriate to seek shelter and to inhibit foraging behaviour. Thus, any meaningful assessment of the adaptive value of behaviour can never be done in 'absolute' terms but only in relation to prevailing circumstances.

Coping strategy

Dingemanse *et al.* (2009) make a distinction between the more general variation in behavioural style (animal 'personality' overall) and **coping style** (the suite of behavioural and physiological responses of an individual that characterise its reactions to challenge across a range of stressful situations).

Welfare

Welfare describes an internal state of an individual, as experienced by that individual. This state of welfare is the result of an interplay between the individual's own characteristics and the environmental conditions to which it is exposed and cannot be assumed to be the same for all individuals placed within a given environmental situation. Human determination of an animal's state of welfare is only as good as the observer's perception of the signals that the animal emits. A negative state of welfare is perceptible via reactions that are aimed at changing the existing situation. A neutral or positive state of welfare is perceptible via lack of any reaction or reactions aimed at keeping the existing situation as it is.

The welfare state of an individual represents a function of its adaptive functioning within prevailing environmental circumstances. For social animals, that environment includes other members of the social group or population. A separate assessment of welfare at the group or population level may thus be determined as the adaptive functioning of the group as a whole in response to a given welfare challenge. The adaptive functioning of a group is the result of the characteristics of that group, as well as the environmental conditions to which the group is exposed.

At the individual level we assume that welfare is determind by the animal's ability and freedom to adapt to environmental conditions. But we recognise that individuals may show significant variation in their perception of a given status and their 'decision' about how to respond to that perceived status. Thus we may expect that even under identical environmental conditions, different individuals within a group or population may perceive or experience their welfare status differently.

Welfare status

The term 'welfare status' as used in this book refers to the factual, biological status of an animal or group of animals. As such it describes a biological status that may be bad, good or neutral, but is *per se*, neither morally bad nor good. In our terminology, a welfare **problem** occurs if the adaptive capacities of an individual or group are being exceeded.

Whether or not the welfare status of an individual or group of individuals constitutes a welfare **issue** implies a value judgement by an observer or by society. Throughout, we try to distinguish clearly that the biological status of an animal should be disentangled from the moral dimension that is brought in by a human observer who interprets and values any apparent animal welfare problem.

Positive welfare

Positive (or good) welfare describes the state in which an individual has the freedom adequately to react to the demands of the prevailing environmental circumstances, resulting in a state that the animal itself perceives as positive. In addition, with a growing emphasis on the importance of positive experiences, good animal welfare is not ensured by the mere absence of negative states but requires the presence of positive affective states.

At the group level, positive (or good) state of adaptive functioning describes the state in which a group has the freedom adequately to react to the demands of the prevailing environmental circumstances, resulting in a state that all individuals of that group perceive as optimum or at least satisficing.

Negative welfare

In our view, and as a view increasingly expressed in the wider literature (e.g. Broom, 2006; Korte *et al.*, 2007; Ohl and van der Staay, 2012), negative or bad welfare status describes a state that the animal itself perceives as negative. A brief state of negative welfare may fall within an animal's adaptive capacity and would not necessarily require intervention. Welfare status is more significantly compromised when an animal or a group of animals have insufficient opportunity (freedom) to respond appropriately to a potential welfare 'challenge' through adaptation by changes in its own behaviour (either where environmental challenges exceed the adaptive capacity of the animal or the opportunities available are inadequate to permit the animal effectively to express the appropriate adaptive responses). Negative welfare therefore describes a state that the animal itself perceives as negative but that still lies within the animals' adaptive capacity.

The adaptive functioning of a group is compromised ('welfare status' is negative or bad) when a group of animals have insufficient opportunity (freedom) to respond appropriately to a potential 'challenge' through adaptation by changes in behaviour (where environmental challenges exceed the adaptive capacity of the group as a whole or the opportunities available are inadequate to permit the group to effectively express the appropriate adaptive responses). Negative 'welfare' at the group level therefore describes a state in which distinct individuals may still perceive their own state as positive but that does not allow for adaptive functioning of the group as a whole. In such a situation, it can be expected that the number of group members experiencing 'negative welfare' or 'suffering' will progressively increase over time.

Suffering

Suffering describes the negative emotional experience resulting from being exposed to a persisting or extreme negative state of welfare. Short-termed, negative welfare states such as suffering from hunger and fear serve as triggers for the animal to adapt its behaviour. They therefore serve a function. A brief state of negative welfare may fall within an animal's adaptive capacity and would not necessarily require intervention. If an individual lacks the ability or the opportunity to react appropriately to suffering (for example, by escaping from a frightening situation), a challenge is created that may exceed the adaptive capacity of the individual. In such a case, the situation is one of suffering for which intervention may be required.

Unavoidable suffering

There may be constraints on intervention that are posed purely by practicalities of intervention or mitigation. Effective intervention in the lives of free-ranging or wild animals may simply not be feasible; in a farm context, if livestock animals are kept outdoors (which may itself be warranted in terms of promoting positive welfare in other respects), they may be exposed to extreme weather that may cause transient suffering. Sudden changes of weather are neither predictable nor controllable and occasional suffering is unavoidable when animals are kept outdoors.

Necessary suffering

Other constraints may be posed by anthropocentric aims or human convenience, for example efficiency of animal production in an agricultural context. In such cases, there may be mitigation options that are not implemented for reasons of efficiency or economics. Here, any suffering resulting from non-intervention may be considered avoidable (in theory) but *necessary* (because of distinct subjective/individual human interests).

Acceptable suffering

There may in practice be a distinction between what an individual producer or manager deems necessary or unavoidable and what society as a whole considers acceptable, thus whether some instance assessed as necessary suffering may or may not in fact be *morally acceptable*. This returns us to considerations of what is or is not acceptable to society as a whole. What is acceptable or unacceptable to society is inevitably a value judgement or ethical decision formed within what is the ethical view of contemporary society.

Welfare issue

As above, a welfare issue arises when a welfare status is perceived as a moral problem by an outside observer who considers that the welfare status is unacceptable and needs to be addressed (the animal or animals are in a poor welfare state, or at least a state which might be improved upon).

Self-perception (of an animal)

In effect the 'decision' by any individual animal to accept its current status or to engage in behaviour designed to bring about some change of status must in part be determined by an assessment of physiological condition (hunger, thirst, etc.) but also by an assessment of a sense of 'well-being'. Most commentators now agree that welfare is largely a function of an individual animal's own perception of its internal status. Although 'perception' does describe some cognitive processing of internal information, it does not necessarily imply a process of conscious reflection; rather, self-perception in animals should be understood as a process of relating internal information to external stimuli.

Emotional state

It is clear that emotions play an important role in this assessment and in the performance of adaptive behaviours. There is a growing literature to suggest that much of the function of emotion or emotional status may indeed be to provide a convenient proximate surrogate to reinforce behaviours that are (or were) in some way adaptive, to make performance of these appropriate behaviours in some sense pleasurable or rewarding and thus promote their expression in appropriate circumstances.

When therefore we, or other authors, refer to an animal's emotional reaction to a given situation, we refer quite explicitly to those physiological and behavioural mechanisms which enable that individual to assess its internal and external situation and trigger appropriate adaptive responses.

Maximising versus satisficing

It is perhaps self-evident that if some animal is perceived by the observer to be in a negative welfare status, but has opportunities (correct behavioural repertoire, appropriate environmental conditions) to improve its status, yet fails to take that action, it may simply perceive its own status as satisfactory.

It seems probable that not all individuals necessarily seek to maximise welfare status at any given point, and that 'adequate' may at times be enough. 'Satisficing' is an alternative to maximisation for cases where there are potentially many possible alternative options that cannot be fully evaluated effectively. A decision maker who gives up the idea of obtaining an optimal solution but obtains the optimum he can compute under given time or resource constraints is said to be satisficed.

In this approach, one sets lower bounds for the various objectives that, if attained, will be 'good enough' and then seeks a solution that will exceed these bounds. The satisficer's philosophy is that in real-world problems there are often too many uncertainties and conflicts in values for there to be a realistic probability of being sure of obtaining a maximisation, and thus that it is far more sensible to set out to do 'well enough' (but better than has been done previously).

It is of note here that the welfare status of the **group** may be optimised, while the (apparent) welfare states of its (satisficing) individual members may vary over a considerable range.

Stress

In general, physiological approaches to assessment of welfare status have focused on the concept of measuring levels of stress experienced by individuals based on the belief that if stress increases, welfare decreases. However, there are a number of problems with such an approach. Short-term stress responses are an inevitable part of the process triggering an adaptive response from the animal and thus may be functional in maintaining a longer term positive welfare status. In such analysis, a more relevant measure might be evidence of chronic and 'traumatic' stress (that exceeds the individual's adaptive capacity), something which is not trivial to differentiate by means of physiological measurements from acute stress.

Stressor

When we use the term 'stressor' we do not necessarily imply something which causes 'stress' that exceeds the individual's adaptive capacity, but, in effect, denote any environmental or other force impinging on an individual animal or group – an environmental 'challenge' which has to be met.

INDEX